rspb

MIGRATION HOTSPOTS

The World's Best Bird Migration Sites

Tim Harris

B L O O M S B U R Y

LONDON · NEW DELHI · NEW YORK · SYDNEY

The RSPB speaks out for birds and wildlife, tackling the problems that threaten our environment.
Nature is amazing – help us keep it that way.

If you would like to know more about The RSPB, visit the website at www.rspb.org.uk
or write to: The RSPB, The Lodge, Sandy, Bedfordshire, SG19 2DL; 01767 680551.

First published in 2013

Bloomsbury Publishing Plc, 50 Bedford Square, London WC1B 3DP
Bloomsbury USA, 175 Fifth Avenue, New York, NY 10010

www.bloomsbury.com
www.bloomsburyusa.com

Bloomsbury Publishing, London, New Delhi, New York and Sydney

A CIP catalogue record for this book is available from the British Library
Library of Congress Cataloging-in-Publication Data has been applied for

Commissioning editor: Julie Bailey
Text, Editorial and Design: Windmill Books Ltd.

UK ISBN (print) 978-1-4081-7117-2
ISBN (e-book) 978-1-4081-6003-9

Printed in China

10 9 8 7 6 5 4 3 2 1

FSC
www.fsc.org
MIX
Paper from
responsible sources
FSC® C104723

Cover photo credits
Front cover: Sandhill Cranes by David Tipling
Title page: Barn Swallow by Neil Bowman
Page 4: Eurasian Stone-curlew by Neil Bowman

Contents

Conservation of birds' breeding and wintering habitats is crucial, but for migrants such as Eurasian Stone-curlew (*Burhinus oedicnemus*) this is a wasted effort without protection of the sites they use on passage.

FOREWORD

The author says that this book aims to be a 'celebration of migration'; it easily fulfils its aims. If you want to catch up on how scientists are improving our understanding of how birds find their way across the globe, this is not the book for you. However, if you want to get a feel for one of the most amazing natural phenomena that our planet has to offer, then pick it up – and be amazed. Travel with the author to fantastic places to see birds on the move in their millions and learn what we know about who goes where and when. Many of these most famous sites are spectacular for their huge numbers of birds of prey – over one and a half million have been counted passing Veracruz, Mexico, in a single day. Many smaller birds, however, travel at night and even when flying in the day may be too high to see. Nevertheless 'back-of-the-envelope' calculations indicate that on average 10 million birds must cross each kilometre of the North African coastline during the two-month autumn migration period as they pour unnoticed into Africa for the northern winter.

This book covers many of the most famous spots for watching migration. A bit surprisingly – partly due to the difficulties of getting to some of these places – some have only recently been recognised. Inevitably many of these sites involve migrating birds of prey and storks for these are often by far the most conspicuous – in order to minimise the amount of energy-sapping flapping flight, whenever possible they use updraughts where they can soar without flapping. Such updraughts are often very localised, resulting in spectacular concentrations. However, at other sites such as Falsterbo, Sweden, or Matsalu Bay, Estonia, one can also see large numbers of songbirds, waders and ducks.

Such sites are commoner in the northern than the southern hemisphere because there are large land masses for birds to migrate between. But huge numbers of these 'northern' birds move to warmer climes during the northern winter and so one can see large numbers of migrants in places in Malaysia, China and India; northern wading birds even reach Australian shores in abundance. Some 29 key sites are described – enough to whet anyone's appetite for travel. Each is accompanied by striking photographs. This is a book to relish, to admire the amazing numbers and amazing journeys. Nevertheless, many entries draw attention to threats to the migrants – not always at the sites themselves, which are often protected, but to the birds when they are on their breeding or wintering grounds. All the usual culprits are listed, including hunting, development and pollution. Unfortunately for migrant birds another threat is getting ever larger – the spread of supposedly environmentally friendly wind-farms results in some migrant birds flying across areas which daily become more like some giant mincing machine.

Professor Christopher Perrins
Edward Grey Institute of Field Ornithology, University of Oxford

Introduction

Little did I know it at the time, but it was the appearance of a wanderer from across the Atlantic that sparked a lifelong interest in birds. For some reason lost to me now, I agreed to a friend's suggestion to visit some local marshes after school. I certainly wasn't convinced by his argument that there was a rare gull there and could imagine my response being "So what?" However, I did accompany him and his pair of ancient binoculars – and I have no regrets.

The mudflats around the grazing marsh were smelly, and the rotting hulks of old boats littered the shoreline. There was certainly no interpretive sign to greet us at the gate, just a burnt-out car. The place seemed slightly sinister, but there was something magical about it as well. It was a warmish, sunny late afternoon in April and I still remember as if it was yesterday hearing Skylarks singing and the bubbling call of a Curlew (no 'Eurasian' in those days), things I'd never been aware of before. We found a Curlew to look at, too, and it was just like the picture in my friend's *Observer's Book of Birds* – only the real bird was in colour. Not everything was easy to identify, however, and there was no doubt we were absolute beginners. We even struggled over the identity of some 'fat-winged' black-and-white birds, which we saw fly to the north and later concluded were Lapwings. So it's little wonder we didn't find the Franklin's Gull, a vagrant from North America and a real 'mega'. For me, it mattered not. I was hooked, and a small flock of passage Lapwings and a couple of Curlews were to blame.

The following spring, armed with my own binoculars, a field guide and with a life-list of – well, at least 80 – I embarked on an expedition to Selsey Bill to look for signs of migration. I say 'expedition' because getting there involved two bus journeys and a lot of walking, though I managed to hitch a lift home afterwards. But it was all worth it, with plenty of migrant activity; and I still have the sketch map to show what was moving where. Sand Martins, Swallows, House Martins and Goldfinches came in off the sea; an Arctic Skua, eight Red-breasted Mergansers and plenty of Sandwich and Common Terns were moving east; and 19 Common Eider for some reason flew the 'wrong' way. At Church Norton and along the seawall to Sidlesham Ferry there were singing Blackcaps, Willow Warblers and Chiffchaffs. And to cap it all, when I was almost at Sidlesham, there was a first-winter Little Gull. This was great. Birds weighing just a few grams had crossed continents to get here, and I could actually see them as they arrived! Ever since, I've been fascinated by migration.

The exploitation of food resources

Migration is movement with a purpose. It raises all the issues of why, how, when and where. This isn't the place to describe the origins of migratory behaviour, but it has clearly provided benefits for those species that practise it. Suffice to say the main driver is the need to exploit the best food resources, especially during the breeding season. Crudely put, plentiful food means more young fledged. Most of the world's long-distance terrestrial migrants move between high latitudes in the northern hemisphere and tropical or subtropical regions. The former have an abundance of food, especially insect food, during the long days of summer and the latter have plentiful food all year round, though there is strong competition for it. There are equivalent migrations from high latitudes in the southern hemisphere but on nothing like the same scale simply because they cover a much smaller proportion of our planet's land area – apart from Antarctica, which is almost inhospitable for terrestrial species.

When, where and how

The timing of migration is triggered by changes in day length, as yet not fully understood hormonal changes, and local weather conditions. Of course, conditions may be conducive to migration locally but not so some distance away, and although birds probably have the ability to interpret changes in air pressure (and therefore changes in the weather), they can still run into unexpectedly wet and windy conditions. It is these shocks that result in 'drift', 'falls' or vagrancy.

Some birds migrate mostly at night and others are primarily diurnal but these divisions are not rigid. A flock of Redwings that embarks on a North Sea crossing some time after nightfall may not reach the other side until long after sunrise. Likewise, many generally diurnal migrants will fly at night if they have to. We still don't fully understand how birds navigate, but several mechanisms are used: the sun, moon or stars as 'compasses'; an ability to 'read' the Earth's magnetic field; the skill to interpret the landforms they are flying over; and learned familiarity with terrain that has been visited before.

Epic journeys

Ponder on some of the awesome feats of physical endurance performed by billions of birds. Some Willow Warblers, for example, weighing about the same as two teaspoons of sugar, fly between eastern Siberia and southern Africa and back each year, a distance of nearly 12,000 kilometres each way, the longest migration of any passerine species. At least they have the luxury of being able to stop and refuel en route, unlike those Bar-tailed Godwits that fly non-stop over the Pacific Ocean between New Zealand and Alaska, around 11,000 kilometres.

Even the world's biggest mountains cannot prevent Bar-headed Geese travelling between Tibet and India. These wildfowl fly over the Himalayas and have even been heard over Mount Makalu at 8,500 metres. Whooper Swans have been observed on migration between Iceland and Scotland at 8,200 metres, where the air temperature is -40^{0}C.

Outside the breeding season many species of seabirds spend most of their lives moving around the oceans. The Sooty Shearwaters that we may see passing a Cornish headland in August or September will have journeyed 15,000 kilometres since leaving their breeding colonies around the Falkland Islands. They follow a looping route around the Atlantic Ocean, following the eastern seaboard of the Americas, and then use transatlantic winds to help transfer to the coast of north-west Europe before heading back to the South Atlantic. Their shearing flight means that gale-force winds, rather than buffeting them, propel them at impressive speeds; they are capable of covering 500 kilometres in a day. New Zealand Sooty Shearwaters have been tracked across up to 74,000 kilometres in a year, so a 30-year-old bird could have flown the equivalent of 50 circuits of planet Earth. The 30,000 to 40,000-kilometre Antarctic–Arctic–Antarctic marathon of Arctic Terns is well known; less so is the example of a fledgling of this species ringed on the Farne Islands, Northumberland, and seen near Melbourne, Australia, three months later.

Physical barriers

Migrants are restricted by physical barriers and the absence of food. Despite the altitude records of the previously mentioned Bar-headed Geese and Whooper Swans, most birds will migrate around, rather than over, the biggest mountain ranges. For broad-winged raptors and storks, flapping flight involves using large amounts of energy; they prefer to glide. Gliding flight is assisted by thermals, which form over land on hot days, but not at night and not over the ocean. Wide tracts of water represent obstacles for these species, which is why they converge on the shortest crossings when migrating – places such as the Strait of Messina, the Bosphorus, Falsterbo and Ras Muhammed.

Not all raptors suffer such constraints. Consider for a moment the 22,000-kilometre round-trip of Amur Falcons between their winter domicile in southern Africa and nest-sites in north-east China and the far east of Siberia. They can't afford to reach China too early since it will be frozen and they will starve. But neither can they afford to get there too late and miss out on the best territories. These falcons migrate through eastern Africa in double-quick time in spring then head out directly across the Indian Ocean to save time, a journey that takes them over 2,500–3,100 kilometres of ocean; they fly non-stop for two or three days and nights. Then, it's across India and South-east Asia, some birds making another sea crossing, over the Gulf of Bo Hai, to make landfall in north-east China. Extraordinarily risky as it is, the Amur Falcons' epic journey must be worth it for the benefits it brings of improved breeding success.

Migrating Barnacle Geese (*Branta leucopsis*) arrive at Matsalu Bay, Estonia.

Human intervention

Human activity, however, can completely upset the balance. In autumn this falcon faces the additional hazard of hunters in north-east India, who are thought to be responsible for netting, killing and skinning an estimated 120,000–140,000 birds each autumn for food. Worse, in some parts of the world, notably parts of the Mediterranean, similar carnage is carried out not to fill empty stomachs but in the name of 'sport'. Spoon-billed Sandpiper is a Siberian tundra-breeding species whose numbers have fallen to desperately small levels for reasons that are not fully understood. What we do know, however, is that the ongoing development of much of the coastline of East Asia has compromised the intertidal zones that they depend on as stopover sites when migrating to and from South-east Asia.

In May 2011, while birding in Beidaihe, China, I was privileged to be involved in the discovery of a bird that wasn't even on my radar when I'd arrived. To be honest, it would almost certainly have gone unidentified had Mark Andrews (of the Oriental Bird Club) and a couple of sharp photographers not been at the scene. It was only later that the magnitude of this find – a Streaked Reed Warbler – dawned on me. It is a long-distance migrant whose breeding grounds are still unknown; observations in its winter domicile on Luzon, in the Philippines, have dried up; and there were five sightings of migrants in the period 2008–2011. That's five. In the world. It is far from inconceivable that this migrant *Acrocephalus* could become extinct without us even knowing where it breeds. Sadly, it is not alone, and although BirdLife International and the conservation bodies have done a superb job of publicising many of those species that are on the brink, what is so poignant about *Acrocephalus sorghophilus* is that few people have even heard of it.

An Amur Falcon (*Falco amurensis*) arrives on the northern coast of China in May after a migratory flight that has brought it from southern Africa.

The migration flyways

Eight major flyways form conduits for the huge transfer of continental birds from the northern hemisphere to the tropics and beyond, but these avian super-highways are not rigid; rather, they have elastic boundaries. Some species use the whole length of their respective flyway, others just part of it; some switch from one flyway to another. Many species use staging sites for resting and refuelling before the next stage of the journey: intertidal flats and wetlands perform this role for waders; lakes, estuaries and pasture offer the same opportunities for wildfowl.

Pacific Americas

More than 300 species use the Pacific Americas Flyway, which stretches from the tundra of the far east of Siberia through Alaska, British Colombia and the western seaboard of the United States, Central and South America to Tierra del Fuego. Millions of wildfowl, waders, passerines and, to a lesser extent, raptors use this route. Notable stopover sites include the Copper River Delta, Alaska, where huge numbers of Western Sandpiper and Dunlin gather; the Fraser River estuary, Canada, where waders stop off and American Wigeon spend the winter; Panama's Upper Bay, which is used by more than a million northern-breeding waders; and Ecuador's Humedales de Pacoa lagoons, the wintering area for tens

Hundreds of thousands of Sandhill Cranes (*Grus canadensis*) congregate at the Platte River, Nebraska, on their spring migration. This is a key staging post on the Central Americas Flyway.

of thousands of Wilson's Phalaropes. Millions of passerines and near-passerines also use the route. Olive-sided Flycatcher migrates from Alaskan forests to the Andean foothills of Bolivia, the longest journey of any tyrant-flycatcher. Some tiny Rufous Hummingbirds fly 9,500 kilometres between Alaska, where they breed, and Mexico, some of them passing through Point Reyes, California.

Central Americas

More than 380 species migrate along at least part of the Central Americas Flyway, which extends 14,000 kilometres from Arctic Canada to the southern tip of Argentina and overlaps the other two American flyways. Bobolinks fly 9,000 kilometres from the Canadian prairies to the pampas of Argentina, where up to 500,000 congregate at the San Javier rice paddies every winter. Another impressive migrant is Buff-breasted Sandpiper, a tundra breeder with a loop migration route: it flies north across the Great Plains on its spring passage from southern South America to breed in Arctic Canada, but in autumn it follows a more easterly route. Not all the migrations are North America–South America epics. Golden-cheeked Warbler, for example, moves between the juniper–oak forests of western Texas (where it breeds) and the highlands of Guatemala, El Salvador and Nicaragua. For a bird weighing no more than 15 grams that is still quite some feat of endurance. One of the best-known travellers on this flyway is Snow Goose, not least because of the spectacular numbers which migrate together; the Delta Marsh staging area in Canada welcomes more than a million of these birds in autumn. The flyway is also used by so-called austral migrants, those that breed in the southern hemisphere in the austral summer and winter in the tropics. One example of these is Crowned Slaty Flycatcher; it spends the winter in the Amazon Basin after breeding much further south.

Atlantic Americas

The Atlantic Americas Flyway, used by about 400 species, also links arctic Canada with the southern hemisphere but follows a more easterly trajectory – east of the Great Lakes, primarily

Classified as Vulnerable, Grey-sided Thrush (*Turdus feae*) is believed to number fewer than 10,000 adults. It breeds in northern China and winters in Myanmar and Thailand. For limited-range species such as this, knowledge of migration behaviour, as well as breeding and wintering habitats, could be crucial for their survival.

over the Caribbean via Florida, Cuba, the Yucatan Peninsula and Jamaica, then through the eastern side of South America. If distance is the main criterion, Red Knot is the champion of this route, charting a 30,000-kilometre round-trip. Another star is Blackpoll Warbler, which undertakes a 3,000-kilometre ocean crossing between Nova Scotia and the Lesser Antilles. This flight, which cannot be broken for obvious reasons, lasts for up to three days. It is no wonder that many Blackpolls get caught up in vigorous depressions and end up on the Azores, or even Scilly. The majority of small birds opt for the safer option of short sea crossings: Florida, Cuba and the Yucatan Peninsula; or Florida, Cuba, Jamaica and then to the Venezuelan coast. Dramatic visible migration along this route can be seen at Point Pelee and Long Point (where the flyway overlaps with the Central Americas Flyway) and Cape May, in New Jersey.

Palearctic–African migrants

The three flyways that connect the Palearctic and Africa together make up the world's largest migratory system. No one knows the true numbers involved but the best estimates suggest 2 billion passerines and near-passerines, 2.5 million wildfowl, 2 million raptors and millions of waders drawn from all around the Arctic and much of Eurasia and wintering throughout Africa.

Around 300 species use the East Atlantic Flyway, which carries Arctic-breeding wildfowl and waders (from Baffin Island and Greenland to Siberia) to western Europe, the Mediterrean and Africa. It also provides a link for billions of passerines and millions of waders, storks and raptors between the forests, plains and wetlands of northern and western Europe and those of the Congo Basin, Sahel and Mauritania. For storks and many raptors the Mediterranean presents a major obstacle, which explains why Tarifa and Messina are such important convergence zones for these birds – and such good places to see them. Many long-distance passerine migrants use the route, including Northern Wheatears that winter in sub-Saharan Africa and breed in Greenland and north-east Canada; they have to cross the Sahara Desert, the Mediterranean Sea and finally the North Atlantic for the benefit of long hours of summer daylight and abundant insects with which to feed their young. Others travel relatively short distances; for example, Blackcaps breeding in the UK spend the winter around the western Mediterranean, while some central European birds come to the UK. Russian wildfowl make longer journeys, abandoning the frozen wastes that their breeding range becomes for the less harsh estuaries and saltmarshes of the UK and north-west Europe. Aquatic Warbler, Europe's rarest migratory songbird, uses this route.

Mediterranean–Black Sea

The Mediterranean and Black Sea Flyway draws birds from the tundra of Russia and western Siberia; and the forests, plains and wetlands of much of eastern and central Europe and the eastern and central Mediterranean. One of its most significant features is the number of major convergences of migrating raptors:

Messina (Italy), the Bosphorus (Turkey), Batumi (Georgia), the northern valleys of Israel and Ras Muhammed (Egypt) to name but a few. Eilat (Israel), arguably the most exciting migration site on the planet, is an important point on the flyway, which is used by about 300 species. Tragically, one of them – Slender-billed Curlew, which once migrated between Morocco and Russia – is probably now extinct.

East Asia–Africa

The third component of the Palearctic–African system is the East Asia–Africa Flyway, linking points as far as Kamchatka with eastern and southern Africa and used by more than 300 species, four of them critically endangered. One, Sociable Lapwing, moves between Kazakhstan and Sudan. The flyway is also used by some notably long-distance migrants, including Siberian-breeding Northern Wheatears, *yakutensis* Willow Warblers and Amur Falcons. One of the most important staging areas is the Korgalzhyn lake complex in Kazakhstan, which attracts hundreds of thousands of wildfowl and waders in spring and autumn.

Central Asia–South Asia

Another 300 or so species use the Central Asia–South Asia Flyway, which links the Siberian tundra and taiga and huge areas of Central Asian steppe and semi-desert with the wetlands, plains and forests of India, Pakistan and Iran. Birds in the eastern half of the flyway face the huge obstacle of the Himalayas; most have to go around these formidable mountains, though there are notable exceptions. To the west, Chokpak Pass (Kazakhstan) is one of the best-known north–south routes through the huge mountains of Central Asia; it enables millions of migrants to reach the vast steppes and forests beyond. Visible migration can be impressive in spring and autumn for groups as diverse as herons, raptors, waders, bee-eaters and passerines.

Didric Cuckoo (*Chrysococcyx caprius*) is a brood parasite and short-distance seasonal migrant in tropical and sub-tropical Africa, moving with the rains. This behaviour enables the cuckoos to take advantage of the caterpillars which make up the bulk of their diet.

Tracking migration routes using new technology

For more than a century, much of our knowledge of migration patterns has been gleaned from ringing programmes. Birds are caught in mist-nets or Heligoland traps, a ring is placed on one leg and, should the bird be retrapped (or found dead) elsewhere, conclusions can be drawn about the route followed (as well as the bird's age and biometrics). The disadvantage is that the chances of an individual migrant being retrapped are slim, so large numbers have to be processed and even then recoveries are unusual. With the advent of satellite tracking technology, it became possible to fit small transmitters to birds to trace their movements and provide accurate information not just on their end destination but the routes followed. The disadvantages are that this is not suitable for small birds and it is expensive.

With concerns growing about the decline of the breeding population of Common Cuckoo, the British Trust for Ornithology (BTO) launched a satellite-tracking programme to shed light on problems that these long-distance migrants face on their wintering grounds or on passage. The early results were fascinating. They showed that British-breeding birds do not all follow the same route back to the Congo Basin in late summer and autumn. One made the short crossing over the Strait of Gibraltar before skirting around the west of the Sahara Desert and finally taking a route to the south-east. Others flew the length of Italy before making a much longer crossing of the Mediterranean into Libya and then directly across the Sahara. And two birds made landfall on the north coast of Egypt to follow a much more easterly course.

It was also demonstrated that the birds did not make regular progress on their way south. They spent July and August wandering around southern Europe before flying rapidly across the Med and the Sahara; then their progress slowed again once they reached the Sahel. Individuals did not stick to the same route. One of the Cuckoos, called Lyster, flew around the west of the Sahara in autumn 2011 but followed a more direct route over the desert in spring 2012 and then used this course again the following autumn.

Bohemian Waxwing (*Bombycilla garrulus*) breeds in taiga forests in North American and Eurasia, moving south in winter. In irruption years tens of thousands move further west than usual, even reaching the British Isles.

To the south, the Rann of Kutch (India) is an important migratory end-point and staging site, attracting thousands of Ruddy Shelduck, Northern Shoveler, Northern Pintail, Common Crane, Little Stint and Ruff.

East Asian–Australasian

Finally, there is the East Asian–Australasian Flyway, connecting eastern Siberia, China, Korea and Japan with South-east Asia and Australia. Among the 500 species that use the flyway are seven critically endangered birds, including Spoon-billed Sandpiper and Black-faced Spoonbill, both of which can be seen at the great waterbird staging site of Mai Po (Hong Kong). A staggering 50 million waterbirds are thought to use the route, including almost three million Oriental Pratincoles, one million Marsh Sandpipers and four million other waders. Many wildfowl and waders migrate between the northern hemisphere and Australia, but few land birds do, Oriental Cuckoo, White-throated Needletail and Fork-tailed Swift being three notable exceptions.

Irruptive species, altitudinal migrants and nomads

Many birds undertake movements – some long, some short – that do not fit the flyways template. In Africa many short-distance migrants and nomadic species move to where the rains are, guaranteeing better supplies of food. Birds living in drought-prone areas in Australia and parts of South Asia abandon their usual range in search of water. Other species descend from mountains to lower altitudes in winter to escape starvation, a form of short-distance migration – Capped Wheatear in Kenya is a good example.

Less predictable are irruptive movements. If the population of a usually sedentary species outstrips its food supply, such as happens sometimes with some species of owls, Spotted Nutcracker, Bohemian Waxwing and Pine Grosbeak, for example, the birds will 'irrupt' or move – sometimes long distances

– away from their usual range in search of food. Unusually harsh weather forces birds to make movements they would not usually make. For example, a big freeze in continental Europe will force millions of ducks and herons to seek open water to the south and west.

Selecting the sites

The scope of this books means that it has only been possible to cover a selection of the world's greatest migration hotspots so I debated with myself long and hard over which ones to include. I felt it was important to have at least one location on each of the eight major flyways, while maintaining a balance between spectacular wildfowl flyways, raptor bottlenecks, passerine visible migration sites, rarity traps and more 'subtle' but equally valuable locations, such as those where nocturnal migrants turn up almost unnoticed (Fraser's Hill, for example). Consequently, there are some well-known absentees, including Pennsylvania's Hawk Mountain and the Bosphorus. The inclusion of the 'new' locations of Batumi (Georgia) and Chumphon (Thailand) ensures that raptors get good coverage. I have also included a couple of 'leftfield' locations (Point Reyes and Moreton Bay) where it is possible to experience the migration of seabirds, wildfowl, waders and passerines, along with the chance of a rarity.

This book aims to be a celebration of bird migration, and above all, I hope it inspires. If just a few of those who leaf through its pages are convinced that more needs to be done to save the world's migratory birds, it will have served its purpose.

Conservation and education

Migrants face a multitude of man-made hazards, from coastal development to pollution, long-line fishing to overhead power lines and wind farms. International bodies such as BirdLife International and WWF, and their national partners, have done tremendous work to push protection for sites and habitats used by migrants higher up the agenda. However, the pressures remain; indeed, they are growing in many parts of the world. Even small voluntary organisations can make a difference, though. The trapping and shooting of migrant birds – especially, though not exclusively, birds of prey – is a major problem. In some poor regions they have long been seen as a valuable source of protein, a natural gift. This was one of the issues faced by the researchers involved in the Batumi Raptor Count project in Georgia. They took the decision to engage with, rather than confront, the community which has traditionally shot raptors during the autumn passage. Volunteers have described the finer points of identification, discussed the birds' ecology and explained how they are a unique resource for local people. Project members have set up classes for children, produced a booklet for local people about the ecology of migration and encouraged ecotourism, which will help the local economy – but only if the birds are still there.

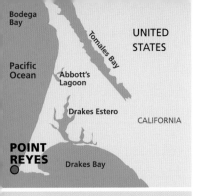

Bodega Bay

Tomales Bay

UNITED STATES

Pacific Ocean

Abbott's Lagoon

Drakes Estero

CALIFORNIA

POINT REYES

Drakes Bay

- **LOCATION**
  Coast of northern California, United States

- **FLYWAY**
  Pacific Americas

- **SPRING**
  Early migrants include Barn Swallow, Rufous and Allen's Hummingbirds, Red Knot and other waders, Caspian Tern and Pacific-slope Flycatcher. Later, Warbling Vireo, Western Kingbird, Western Tanager and Fox Sparrows move north. Offshore, large numbers of scoter, divers, gulls and terns.

- **AUTUMN**
  July is good for albatrosses in offshore waters. In August and September these are largely replaced by a variety of petrels, shearwaters and storm-petrels. Waders stream along the coast, northern-breeding wildfowl appear offshore and on fresh water and passerines move south. Rare songbirds are often found in October.

- **KEY SITES**
  Pelagic trips from Bodega Bay; Point Reyes National Seashore, including Point Reyes Lighthouse, the Fish Docks and Abbott's Lagoon; Bodega Head; Hawk Hill.

Point Reyes, USA

It is August and the engines are driving the boat steadily west on a gently rolling ocean blanketed by chilly fog. Suddenly the visibility increases and the sense of anticipation is tangible. Binoculars focus on the horizon because here the storm-petrels are Ashy and the shearwaters are Buller's. Welcome to northern California!

Point Reyes has the lot: many mind-blowing seabirds passing offshore on their way through the eastern Pacific; thousands upon thousands of regular wildfowl, wader and passerine migrants in transit along the Pacific Americas Flyway in spring and autumn; and an enviable list of strays from eastern North America and even Asia.

The California Current

What makes the place so special? Hammerhead-shaped Point Reyes juts 15 kilometres into the cold, bird-rich California Current. Behind its cliffs is a triangular peninsula whose broad range of habitats – coastal scrub, coniferous forest, dunes, sheltered inlets with intertidal mud, dairy farmland, brackish and freshwater lagoons with reedbeds – exert a strong pull on migrants and breeding birds alike. The peninsula's bird-list is almost 500-strong and includes 54 species of wader, 25 tyrant-flycatchers and 46 wood-warblers and tanagers.

However, offshore is arguably where the real excitement is. A few kilometres out, a submarine seamount called the Cordell Bank thrusts from the continental shelf to within 35 metres of the surface of the Pacific, then plunges to the depths of the Bodega Canyon to the north and the abyssal plain to the west. The California Current, which sweeps down from offshore Canada, collides with the seamount and brings an extraordinarily rich plankton fauna close to the surface. In turn, this feeds millions of young fish and the result is a feast for seabirds, both local breeders and those in circulation around the Pacific. Seawatching from land can be excellent but it is best to get out where the real action is. More than 30 years of pelagics from Bodega Bay have produced records of thousands of albatrosses, petrels, shearwaters, storm-petrels, murrelets and auklets, not to mention gulls, terns, skuas and phalaropes. Among the seabird masses have been picked out the real gems of pelagic legend, including Shy, Short-tailed and Light-mantled Albatrosses, and Manx and Streaked Shearwaters. The petrel list is particularly impressive, including Parkinson's, Great-winged, Cook's, Stejneger's, Mottled and Murphy's. Further south, the Half Moon Bay and Monterey pelagics also explore these exciting waters and have their own blend of seabird specialities.

Point Reyes lighthouse is at the tip of the peninsula of the same name. Its light is a point of attraction for migrants that have found themselves out at sea. Fog is commonplace over the cold California Current, a few kilometres offshore.

Spring migration

Spring migration starts in late February when Allen's Hummingbirds return from their winter sojourn in the south of the state and Violet-green Swallows pass through on their way to British Columbia and Alaska. Early March arrivals include Rufous Hummingbirds and Barn Swallows from Mexico, then Red Knot, Caspian Terns, Northern Rough-winged and Cliff Swallows and Bullock's Orioles. Pacific-slope Flycatchers are the first of the empids to pass through and Wilson's Warblers claim the same accolade for the wood-warblers. Offshore, divers and scoter pass north.

The main rush of spring's passerine and near-passerine migration occurs in April, with Warbling Vireos followed by Vaux's Swifts, Ash-throated Flycatchers, Western Kingbirds, Cassin's Vireos, Swainson's Thrushes, Yellow and Black-throated Grey Warblers, Western Tanagers, Golden-crowned, Chipping, Fox and Lincoln's Sparrows, Black-headed Grosbeaks and many more. Later in the month, Olive-sided Flycatchers, Western Wood-Pewees, Grasshopper Sparrows and Lazuli Buntings are on the move. In May there are more northbound scoter and divers, a chance of albatrosses and the first Arctic Terns and Sabine's Gulls, though the latter are more likely in autumn.

Pink-footed Shearwater (*Puffinus creatopus*) is a transequatorial migrant with breeding grounds on islands off the Chilean coast. It appears off northern California between April and November, often feeding in loose association with Buller's and Sooty Shearwaters.

Summer and autumn

By June, migrant waders will almost all have passed through and passerines will either have moved on or settled down to breed. However, offshore there will be albatrosses and an increase in the number of Pink-footed and Sooty Shearwaters. In July, Laysan and Black-footed Albatrosses are regularly seen on Cordell Bank pelagic trips, with the latter much more frequent. By August, the region is under assault from two sources: seabirds and waders. While albatross numbers decline (almost to a trickle in late August), Sooty Shearwaters are on the increase and Sabine's Gulls and Arctic Terns begin to return. The first autumn groups of Ashy Storm-petrels and Buller's Shearwaters are seen at sea, and boat trips from Bodega Bay regularly discover tubenose megas at this time. Meanwhile, Greater and Lesser Yellowlegs, Wandering Tattlers, Western, Semipalmated, Spotted and Least Sandpipers, Short-billed and Long-billed

Dowitchers, Black and Ruddy Turnstones, Red Knot, Sanderling and Surfbirds are on the move, appearing on exposed coastal mud and sand or near-coastal wetlands. Abbott's Lagoon, at the northern end of the peninsular, is a good site to search, as is Bodega Head. Baird's and Buff-breasted Sandpipers are also possible. Wildfowl, American White Pelicans, Red-necked and Grey Phalaropes, Arctic Skuas, Bonaparte's Gulls and Common Terns also migrate just offshore. Townsend's Warblers, which have bred in the coniferous forests of the Pacific North-west, reappear from as early as mid-August. Raptor migration is never truly spectacular, but can be interesting at Hawk Hill, where 19 species pass through between mid-August and early December.

Autumn seabirds, wildfowl and passerines

September is the best month for seabird migration. Four species of storm-petrel are possible on a Bodega pelagic (Wilson's, Fork-tailed, Ashy and Black). After north-westerly winds there is a chance of seeing Fork-tailed from land. Further south, off Half Moon Bay, Least and Leach's Storm-petrels can be added to that list. Sabine's Gulls and Long-tailed Skuas pass through in the first half of the month but then decline, and large numbers of Arctic Terns flood south. South Polar Skuas increase in mid-month, while Pomarine and Arctic Skuas are common. Sooty, Pink-footed, Buller's, Black-vented, Flesh-footed and even an outside chance of Streaked and Manx Shearwaters are either likely or at least possible offshore.

Although a nice diversity of petrel species has been accumulated over the years, any petrel sighting is a major event – in contrast to storm-petrels, which can be plentiful.

For land-lubbers, typical September returnees include ducks aplenty, Canvasback, Ring-necked Duck, Bufflehead and Common Goldeneye among their ranks. This month is also the most exciting for landbird migrants. Say's Phoebes, American Pipits, Ruby-crowned Kinglets, Varied Thrushes, Cedar Waxwings and Fox Sparrows return on their way south. Hawk Hill is the best site in California for Broad-winged Hawks; small numbers are regular in late September.

A few Short-tailed Shearwaters are present offshore in October, with bigger numbers of Pomarine and Arctic Skuas; South Polar Skuas peak early in month. The number and variety of storm-petrels peaks, Glaucous-winged and Mew Gulls reappear

Rufous Hummingbird (*Selasphorus rufus*) is one of the earliest spring migrants, arriving in the Point Reyes area in March *en route* to the far north-west of the United States and British Columbia. This bird is a male.

Surfbirds (*Aphriza virgata*) breed in Alaska but winter as far south as the Pacific coast of Chile. In autumn, adults move south along the California coast in July and August, with juveniles following a little later, in August and September.

and Xantus's and Craveri's Murrelets are a possibility. In November, the first Yellow-billed Divers appear within large offshore flocks of Pacific Divers. In October and early November there is the real prospect of a rare eastern North American passerine, or even a vagrant from Asia.

Bodega Bay pelagics explore the waters over the Cordell Bank and Bodega Canyon between late August and late October. Trips into the California Current from Half Moon Bay are organised from late July through to late November. They offer the best chance of seeing Cook's Petrels in late July and early August, while Hawaiian Petrel is a possibility in late August and storm-petrels really come into their own in September and October. South again, the Monterey pelagics run from late August to late October.

The lighthouse and Abbott's Lagoon

Back on solid California rock, Point Reyes Lighthouse is one of dozens of birding locations in Point Reyes National Seashore – and the one with the best rarity pedigree. It stands at the south end of Sir Francis Drake Boulevard. Rare birds have been found everywhere between the lighthouse and its car park, but especially in the cypresses and around the staff housing about 1 kilometre from the car park. The dark area under a particularly low-hanging Monterey cypress bough (at the far end of the first cypress group) was named The Oven by local birder Rich Stallcup after he found his 12th Ovenbird there. It is definitely worth checking in autumn. Scarcities and rarities found nearby have included most of the 'eastern' wood-warblers; Tennessee, Chestnut-sided, Magnolia, Black-and-white, Black-throated Blue, Palm and Blackpoll Warblers are annual. Asian 'megas' have included Dusky Warbler, Red-throated Pipit, Oriental Skylark, Olive-backed Pipit and Eastern Yellow Wagtail.

Grassy areas attract flocks of migrant sparrows, and flowers provide food for passage hummingbirds. And, of course, the seawatching can be productive.

About 4 kilometres away, at the eastern end of the 'hammerhead' that makes up the tip of the peninsula, are the fish docks. Here, the area around the Chimney Rock Trailhead is also good for migrant passerines, typically in the cypresses around the buildings and in the pines towards Chimney Rock itself. Again, a who's who of American wood-warblers has featured here. As at the lighthouse, October is the best month for rarity hunting.

Also worth a visit are Abbott's Lagoon (a few kilometres north of Point Reyes Lighthouse) for wildfowl, rails and waders; the area around Bear Valley visitor centre (near Olema, to the east) for passerines; and, to the south, Hawk Hill (just off Conzelman Road, north of the Golden Gate Bridge), where the Golden Gate Raptor Observatory holds its annual raptor watch between August and early December.

The Cordell Bank pelagic trips depart from the small town of Bodega Bay. Close to the town are several sites to check for migrants, including rarities. Just out of town to the south is the small peninsula of Bodega Head, which shelters the harbour. Near the tip are scrubby Owl Canyon and the freshwater pool–marsh combo known locally as Hole in the Head. In autumn 2011, for example, birders found American Redstart and MacGillivray's Warblers in late August; Prairie, Tennessee, Nashville and Hermit Warblers and Indigo Bunting in September; and Black-and-White, Palm, Nashville and Chestnut-sided Warblers in October. Well worth a look.

The small breeding population of Varied Thrush (*Ixoreus naevius*) in northern California is supplemented by birds from further north in winter. A few strays turn up at Point Reyes, with October being the best month to look for them.

CANADA

ONTARIO

Lake Huron

Toronto. Lake Ontario

POINT PELEE ○ Lake Erie

USA

- **LOCATION**
  North shore of Lake Erie, Ontario, Canada

- **FLYWAY**
  Atlantic and Central Americas

- **SPRING**
  Shore Larks pass through in February, followed by wildfowl and Red-winged Blackbirds in early March. Waders, raptors, gulls and terns increase during early April. May is the peak month, with sparrows, wood-warblers, flycatchers, Ruby-throated Hummingbirds and vireos in large numbers, and thrushes and tanagers also, especially between 10–20 May. Red Knot pass through in late May.

- **AUTUMN**
  The first shorebirds return in late June. Wood-warblers are on the move by August. In September, more warblers, sparrows, raptors, shorebirds and the first wildfowl appear. Raptor passage is often heavy at Holiday Beach in October.

- **KEY SITES**
  Point Pelee National Park, Hillman Marsh, Holiday Beach Conservation Area, Long Point.

Point Pelee, Canada

Without question the best location in inland North America to observe the northward migration of songbirds in spring, Point Pelee is particularly blessed with its diversity of wood-warblers. Up to 39 species of this colourful group have been recorded in a single spring, 34 in a single day.

Despite its richly deserved reputation as the 'warbler capital of North America', at Point Pelee the stream of birds in spring is not a steady flow from the south. The birds arrive in intermittent waves. In some years these are well marked but in others the fluctuations in numbers and variety are so meagre that a 'bird wave' is difficult to detect. What everyone hopes for in spring is a major arrival and a large-scale grounding (or 'fallout') of migrants – an event such as that of 9–12 May 1952 when there were 1,000 Black-and-White Warblers and 20,000 White-throated Sparrows. Or like 15 May 1978 where there were 80 Yellow-billed Cuckoos in the Tip area of the park alone.

Sadly, breeding populations of most North American forest species have fallen (in some cases dramatically) since then so there will probably not be a recurrence of those numbers. However, that does not detract from the anticipation of encountering a

Cooper's Hawk (*Accipiter cooperii*). In autumn this species follows the Lake Erie coast until being funnelled along the Pelee peninsula, from where the lake crossing is shorter. Holiday Beach is a great place to look for this species, especially in early October.

The pulling power of Point Pelee is best seen from the air. It protrudes 19 kilometres (12 miles) into Lake Erie – like a finger pointing at the United States – thus significantly reducing the distance of water that migrant birds have to cross in spring and autumn.

major fall, and the excitement of seeing eight species of spring-bright wood-warblers in the same tree is certainly a tough act to follow.

An epic movement

In the northern summer, plentiful hours of daylight means a huge amount of insect food is available for birds to raise young. The opposite is the case in winter, when the nights close in, temperatures fall and there is little or no insect food. Only the hardiest species can survive those conditions. This is why an estimated 80 per cent of taiga-breeding species fly south in autumn to the southern United States, the Caribbean or South America. Most of the continent's wildfowl and sparrows move within temperate North America, but millions of birds of prey, waders, swallows, tyrant flycatchers, thrushes, vireos and wood-warblers fly to Central and South America. This represents an exodus on an epic scale. An estimated 5 billion land birds migrate from their North American breeding grounds to winter further south. Many millions fly over the Great Lakes area. The migrants use several different flyways, though the distinctions between them are not hard and fast. Some migrate south-east to join the Atlantic Flyway. Others take a more south-westerly route to join the Central Americas Flyway.

Although millions of birds fly over the Great Lakes, human eyes only see a small proportion of them: birds that have stopped off to rest and feed; waterbirds tracking along the Lake Erie shoreline; and large birds, especially raptors, drifting overhead. Weather conditions play a big part in the grounding of migrants at Pelee. In spring, when a warm weather front advancing from the south or south-east meets colder air to the north or north-west, two situations will cause migrating birds to descend. If the fronts meet at ground level over the north shore of freshwater Lake Erie, heavy rain will force migrants down. They will seek cover at the first available landfall, and as Point Pelee protrudes several kilometres south into the lake, many of them pitch down there. Alternatively, if the warm and cold air converge at a higher altitude, and if the birds are travelling north in the warm sector, the front will push them higher and higher. They are forced to descend or risk being carried to an altitude outside of their comfort zone. Pitching into Lake Erie is not an option (although some must do this), so they fly to the nearest landfall. While this 'fallout' may take place on a broad front, Point Pelee's geography means that it benefits disproportionately.

The birds

The total number of bird species recorded in the small area of Point Pelee National Park is 385, of which at least 340 have been logged during the spring migration period. The bread-and-butter species are those that spend the winter in South or Central America, or the southern United States, and breed in large numbers in Ontario or further north. Most major groups are represented: wildfowl, herons, rails, waders, gulls and terns, skuas, cuckoos, swallows, tyrant flycatchers, thrushes, vireos, wood-warblers, tanagers, finches and sparrows. One advantage of birding Pelee in the spring is that – because of its northerly latitude – most arboreal migrants turn up before the trees have all their leaves, making observation that much easier.

What makes Pelee so good? The obvious factor is the lure of its south-spiking peninsula, thrusting out into Lake Erie, and plenty of deciduous woodland from near the Tip to the base of the peninsula holds newly arrived birds for at least a few hours. They can filter north while remaining in cover. The existence of a wetland with muddy and reedy fringes just beyond the woods adds additional habitat variety.

The spring rush

The passerines – starting with the first arrivals of spring – include Shore Lark (from the end of February), American Robin, Red-winged Blackbird, Common Grackle, Eastern Meadowlark, Rusty Blackbird, Brown-headed Cowbird, Eastern Phoebe, Winter Wren, Hermit Thrush, Fox Sparrow, Brown Thrasher, Wood Thrush, Solitary Vireo and Red-breasted Nuthatch. The earliest wood-warblers – Yellow-rumped and Pine Warblers and Northern Waterthrush – appear in early April and by the end of the month the variety is greater. At

An adult male Blackburnian Warbler (*Setophaga fusca*). This can be a common sight on migration at Pelee in May – if conditions are right. This species winters in forests in southern Central America and northern South America and breeds in mature coniferous and mixed forest in eastern North America.

this time sparrow arrivals should include Grasshopper and Vesper, and several species of vireos will have put in an appearance.

Pelee is at its most exciting in the middle of May, when wood-warblers predominate and up to 30 species are possible in a morning on the best days – with a combination of sustained effort and luck! 'Tough' birds such as Mourning and Connecticut Warblers could be found and the potential for rarities is immense. Southern 'overshoots' – Yellow-throated, Worm-eating, Kentucky and Hooded Warblers and Louisiana Waterthrush – are regular and Kirtland's Warbler is now a real possibility. Even Black-throated Grey Warbler, a west coast species, has turned up. Blue Grosbeak, Lark Sparrow and Scissor-tailed Flycatcher were all seen in May 2012. From the end of the third week of the month, the pace slackens although migrant first-year female wood-warblers can still be seen in early June. Remarkably, in contrast to the numbers that use it on passage, only four species of wood-warblers are regular breeders in Point Pelee National Park.

The hummingbird most likely to be seen at Pelee is Ruby-throated (*Archilochus colubris*). It winters south to Panama and arrives on the northern coast of Lake Erie in May. The bird in the photograph is an adult male.

Pelee in autumn

Autumn is a generally a more subdued, but also a more drawn-out, affair. Passerines are less obvious but not because there are fewer of them: many birds are harder to see among the foliage, and the males have lost their brilliant breeding finery. Contact calls replace vigorous song. Raptors are more in evidence than in spring, though Holiday Beach is usually more productive for these than the national park itself. Northern Saw-whet Owls pass through on their way south in October. Rarities are found every autumn. For example, there was a Spotted Towhee (whose nearest breeding is far to the west in Nebraska) in early November 2008.

The Tip, Tilden Woods and Marsh Boardwalk

The visiting birder is certainly well looked after at Pelee. Several well-marked trails run through a variety of habitats. Leaflets and maps are available at the visitor centre, which even has its own gift shop. This is hardly wilderness birding, but it is no less enjoyable for that. It is always worth visiting the Tip (the extreme southern end of the promontory, about 2 kilometres south of the visitor centre) at dawn to see what new arrivals may have come in overnight, what birds are still arriving and what may be moving along the coast. The shrub and tree cover just north of the Tip can be superb. Don't forget to check the shorelines east and west of the Tip for resting or feeding waders. And, of course, there is always a good chance of wildfowl, grebes, divers, gulls and terns passing the point. After making landfall on the peninsula, many birds will gradually filter inland to one of the other areas of cover nearby, so after careful scrutiny of the Tip area, work north, taking care to check the ground for skulking ground-feeders such as Kentucky Warbler, any dense patches of understorey for gems like Mourning and Connecticut Warblers, and the open, grassy areas (notably the Sparrow Field, just north of the Tip) for sparrows. Flycatchers often hawk from the surrounding trees.

The overwhelming majority of Prairie Warblers (*Setophaga discolor*) breed much further south than Pelee, but some 'overshoot' in May and a few breed in parts of southern Ontario. This is an adult male.

Just south of the visitor centre is the Woodland Nature Trail, a 3-kilometre loop through the oldest woodland in the park. It is worth working this – and Tilden Woods, just to the north – thoroughly in search of Red-headed Woodpecker, Yellow-bellied Sapsucker, Ruby-throated Hummingbird, Yellow-billed and Black-billed Cuckoos, tyrant flycatchers, thrushes, vireos and wood-warblers. The Tilden Woods Trail boardwalk passes through some swamp forest and eventually morphs into the Chinquapin Oak Trail, which meanders for 4 kilometres through mixed dry forest. The last main area to check within the national park itself is the Marsh Boardwalk, overlooking a wetland where migrant wildfowl, herons and rails can be seen. There are many places to check and it is not possible to cover all the ground in a single morning. If a lot of birds have arrived it is best to work one area thoroughly rather than everywhere hurriedly.

Scarce birds and rarities

In addition to the 'expected' northern-breeding species, the national park also attracts lesser numbers of birds such as Bewick's Wren, Worm-eating, Hooded and Yellow-throated Warblers, and Le Conte's and Clay-coloured Sparrows (early May); and Bell's Vireo, Blue Grosbeak and Summer Tanager (mid-May). It is just far enough south to get occasional southern rarities: Swallow-tailed Kite, Lesser Nighthawk, Sage Thrasher and Virginia's Warbler are a few examples. Kirtland's Warbler is now almost annual at Pelee, a reflection of its increased breeding success elsewhere.

Kirtland's Warbler

Kirtland's Warbler is now classed as Near Threatened but this actually represents a success story: due to determined conservation efforts it has come back from the brink of extinction. Several factors combined to threaten this yellow and blue-grey *Setophaga*'s very existence: a very small breeding range in the jack pines of northern Michigan; a small winter range in the Bahamas and Turks and Caicos Islands; nest parasitism by Brown-headed Cowbirds; and the many dangers of migrating long distances over land and sea. By 1971 the species was reduced to just 201 pairs. Action was taken in the nick of time. In the breeding range, Brown-headed Cowbirds were eliminated and more jack pines were planted. Numbers increased slowly, then more rapidly. In 2011 there were 1,805 singing males in Michigan, 21 in Wisconsin and two in Ontario, after the first breeding in Canada for more than a century in 2007. The species is now pretty much an annual sighting at Pelee (but far from guaranteed), with the period from 6–16 May being the most likely. Threats still remain: brood parasitism is still an issue; nests are sometimes destroyed by fire; and the loss of Caribbean pines in its winter quarters is a problem.

Hillman Marsh and Holiday Beach

The Essex Region Conservation Authority (ERCA) reserve of Hillman Marsh should not be overlooked later in the day. This wetland is north-east of (and almost adjacent to) Point Pelee National Park. Many ducks, herons, rails, waders, gulls and terns, swallows and other passerines use it as a migration stopover in spring and autumn. For example, 30 species of waders were logged in spring 2005. Fields near the car park are worth searching for Eastern Meadowlarks and Bobolinks in May. At the western end of the marsh is an information centre, 5 kilometres of trails and a tower hide overlooking Hillman Creek. The Onion Fields, just outside the National Park, are often good for waders, especially Grey and American Golden Plovers, Red Knot and Turnstone. Early in the autumn Buff-breasted and Upland Sandpipers are regular visitors.

A few kilometres west of Pelee (and 32 kilometres south of Windsor), along the north shore of Lake Erie, is Holiday Beach Conservation Area. This is another ERCA reserve. Between late August and the end of November, many thousands of raptors pass over. Some years there are also huge numbers of Blue Jays and American Crows, while wildfowl and gulls move just offshore and passerine migration can be impressive. On one late September day in 2009, for example, 21 species of wood-warblers were tallied. However, it is for its birds of prey that Holiday Beach is justifiably renowned, with 15 species regularly noted. The 11-metre Hawk Tower, overlooking Big Creek Marsh, provides an excellent view of raptors and other migrants as they move along the lake shore.

The largest of North America's herons, the Great Blue (*Ardea herodias*), breeds across much of North America but abandons most of Canada when lakes and rivers freeze in autumn. There is a passage of this species through the Pelee area in spring and autumn.

Raptor records

Raptor species peak at different times during the autumn though on one notable day – 23 September 1992 – all 15 of the 'regulars' were seen. In September Bald Eagles, Ospreys and Merlins move through in small numbers. Broad-winged Hawks, Sharp-shinned Hawks and American Kestrels peak in mid-September, sometimes in spectacular numbers: 95,499 Broad-winged Hawks on 15 September 1996 is the record. Cooper's Hawks and Turkey Vultures usually peak in early October, with a phenomenal 20,032 of the latter on one day in 2005. Next it is the turn of Rough-legged Buzzards, Red-shouldered Hawks, Northern Harriers, Golden Eagles and – with the last of the huge numbers – Red-tailed Hawks in mid-November (3,002 on 11 November 1994 is the record). Very rare birds of prey have included a Mississippi Kite on 8 September 2010 and two Swainson's Hawks on 13 October 2009.

Wood Thrush (*Hylocichla mustelina*) has a beautiful flute-like song. This deciduous forest breeder arrives at Pelee around the middle of May and does not leave until September. Sadly, this species has suffered a major population decline in recent decades.

Some of the more notable daily tallies of non-raptors have been 604 Ruby-throated Hummingbirds in mid-September 1997 and 264,410 Blue Jays in late September 2001. The excitement continues in late October with more than 10,000 Cedar Waxwings counted in 1986; 44,000 American Crows and 1,400 American Pipits in 2009; 1,348 Eastern Bluebirds in 1991; and an astonishing 195,000 Shore Larks logged on 5 November 1992. Gulls and wildfowl are late-autumn passage birds at Holiday Beach.

Long Point peninsula

The Long Point peninsula is the longest freshwater sand spit in the world, extending 32 kilometres into Lake Erie and protecting extensive marshes to the north. Unlike Point Pelee it points east rather than south but nonetheless acts as a conveyor belt for migrants and is a great place to watch waterbirds tracking the coast. Daily counts of 70,000–100,000 wildfowl have been made in both spring and autumn, with the numbers of American Black Duck, Canvasback, Red-breasted Merganser, American Wigeon, Ring-necked Duck, Redhead, Lesser Scaup, Greater Scaup and Tundra Swan being globally or nationally significant. Whimbrel, Bonaparte's Gulls and Common Terns also track the coast in large numbers. Incredibly, so far inland, seabird rarities such as Black-capped Petrel and Band-rumped Storm-petrel have been recorded.

The tip of the peninsula can be exceptional in autumn for landbirds, as well. The area's first Sage Thrasher was there on 30 September 2012 and there was a Western Kingbird the following day. And the area around the base of the spit, near Port Rowan, can be superb for passerines, not least wood-warblers. On 11 May 2011 these included a Kirtland's Warbler. Nearby, grackles and Red-winged Blackbirds roost in huge numbers – over 100,000 – during the autumn. Nearly 1 million birds have been ringed at the Long Point Observatory, and it is estimated that 2.4 million birds use the area during spring migration and 7 million in autumn. Long Point's all-time list of bird species is 393-strong, longer even than Pelee's.

American Kestrel (*Falco sparverius*) is North America's smallest and most common falcon. Virtually the whole of the Canadian breeding population abandons the country in autumn.

Cape May, USA

So good it had a wood-warbler named after it, Cape May is the most exciting migration hotspot on North America's eastern seaboard. When conditions are just right in autumn the number of birds passing through can be simply awesome.

Every October, New Jersey birders keep one eye on the birds and the other on the weather forecasts. Any indication that a cold front will be passing during the night, bringing north-westerly winds to the Cape May region, and they will be reaching for their bins and scopes for a pre-dawn raid. The really big falls and the morning flights that follow do not happen very often. But when they do, words are not really enough to describe them. After one such wind change in late October 2010 birders described seeing thousands of birds flying over the beach at Cape May, illuminated by the streetlights long after nightfall. Birds were resting on the street and even in window boxes. Parts of the beach were blanketed by 14 species of sparrows, while Hermit Thrushes carpeted lawns.

When southbound migrants reach the end of the Cape May peninsula in autumn they are faced with a choice: fly out over the sea to reach Delaware and Maryland on the other side, or seek a safer route around Delaware Bay. Some cross, but many nocturnal migrants head north, rather than south, at first light, over Higbee Dike, to find suitable daytime roosts. Cape May's

• LOCATION
Atlantic coast of southern New Jersey, United States

• FLYWAY
Atlantic Americas

• SPRING
Wildfowl, herons, raptors, waders and passerines between early March and mid-June. The first two weeks of May is the best period for variety.

• AUTUMN
Between late July and November, starting with waders, then some species of wood-warblers before raptor migration peaks in late September and the first half of October. Large falls of songbirds possible any time between mid-September and early November. Seawatching can be productive throughout. Rarities sometimes show up as late as early November.

• KEY SITES
Higbee Beach, Cape May Point and Meadows, Stone River Point, Avalon, Heislerville.

• THREATS
The sites are protected.

A male Scarlet Tanager (*Piranga olivacea*) is what it says on the tin. This long-distance migrant winters in Amazonia and the Andes foothills and can be expected at Cape May from early May.

Bay-breasted warbler (*Setophaga castanea*) – this is an adult male in breeding plumage – breeds well to the north of Cape May, where it is more likely to be seen in autumn than spring. Then, immature birds can be confused with immature Blackpoll Warblers (*Setophaga striata*).

expert bird counters are there at dawn to log them on the 'morning flight'. On the morning of Friday 29 October 2010, the legacy of the previous night's arrival must have left memories to last a lifetime. More than 25 passerine species were involved in the morning flight (and many others did not participate), including 73,570 American Robins, 64,640 Yellow-rumped Warblers and 10,000 House Finches. There were also large flocks of Red-winged Blackbirds and American Goldfinches. Meanwhile, the following morning hundreds of American Woodcock left the woods at dawn, the site's first Henslow's Sparrow for many years was found and the raptor watchers counted 455 Sharp-shinned and 115 Red-tailed Hawks, and 105 Northern Harriers. Is it any wonder that Cape May is the most renowned migration hotspot on the east coast of North America?

Cape May lighthouse

Cape May lighthouse and bird observatory sit at the tip of a peninsula that juts south-west across the mouth of Delaware Bay. To the north are New Jersey, the northern states and eastern Canada. To the south are Delaware, Virginia and the southern states. More than 400 species have been recorded at Cape May, beating both Long Point and Point Pelee, a reflection of the coastal position of the site and the fact that it is further south. It hosts land birds moving north or south along the coastal plain and seabirds tracking along the coast, particularly when onshore winds blow them close.

The small town of Cape May is blessed with tracts of woodland, grassy fields, scrub, shallow freshwater pools, small reedbeds, intertidal mud and views over Delaware Bay and the Atlantic Ocean. It has a bird observatory, facilities for visiting birders, a long-standing ringing programme and dedicated bird counters. And like all good coastal migration honeypots it has a lighthouse. What more could you wish for?

The Atlantic Americas Flyway connects the Canadian Arctic archipelago, as far as the extreme north of Greenland, with Tierra del Fuego at the southern tip of South America. Many Arctic breeders use it, along with millions of other birds that breed at lower latitudes in the eastern half of North America. But this flyway is not a clearly defined route. Some birds follow the coast south in autumn to the tip of Florida then cross to South America via the Yucatan Peninsula (Mexico) and Central America. Others prefer to 'island hop' via Cuba and Jamaica to Venezuela. Still more – Cerulean Warblers, for example – follow the Texas coast west to the Rio Grande and then migrate through eastern Mexico.

All the major groups of birds that breed in eastern North America – from wildfowl to waders, raptors to buntings – pass along the Atlantic Americas Flyway and Cape May is well placed to receive them. It has two major advantages over other east-coast sites: being at the southern end of a peninsula, with Delaware Bay beyond, provides a barrier to southward movement in autumn; and the large intertidal zones in the bay attract thousands of waders. Additionally, birds that fly over the eastern Atlantic

An autumn sunset provides the backdrop for Cape May's lighthouse. With cloud thickening in the west, heralding the approach of a cold front and a switch of wind direction, prospects for a fall look promising.

to reach South America bypass the more southerly states of Virginia, the Carolinas and Florida. Many Red Knot and Blackpoll Warblers use this strategy, with the latter crossing the ocean to the Lesser Antilles and South America, a hazardous journey that involves an 80-hour non-stop flight.

Spring migration gets underway

Spring migration begins in March as wildfowl, seabirds, Red-tailed Hawks, Northern Harriers, Piping Plovers and some passerines are on the move. By the third week the migrants' ranks include the first of the wood-warblers, Pine. The flow of wildfowl continues in April, with herons also passing through, as well as some Ospreys, Sharp-shinned Hawks and American Kestrels. Any sudden shower in spring will encourage small birds to pitch down, giving a taste of what is passing, unseen, overhead. Barn, Tree and Northern Rough-winged Swallows, Blue-grey Gnatcatchers, Ruby-crowned and Golden-crowned Kinglets, some wood-warbler species, Brown Creepers and sparrows pass through during the month. Thousands of Northern Gannets, waders, gulls and terns head north, as do American Kestrels, Peregrine Falcons, Merlins, Red-tailed, Cooper's and Broad-winged Hawks. The best conditions for seeing raptors are when there are westerly or north-westerly winds. At other times many move further inland. Raptor migration is never as good in the spring as it is in autumn.

Wood-warblers become more common as April morphs to May. The overshoot birds of prey such as Mississippi and Swallow-tailed Kites, though rare, are a possibility. Raptor passage drops away after mid-May, but waders, vireos, wood-warblers, tanagers and orioles peak in mid-month. June is a month when spring and autumn combine: the last waders and passerines are still tracking north while a few returning waders (possibly failed breeders) pass them on their way back south.

Waders herald the start of autumn

After mid-July autumn migration becomes obvious, albeit only for waders at first. There are more sightings of Greater and Lesser Yellowlegs and Semipalmated Sandpipers, for example. Some of the earliest returning passerines will be Blue-grey Gnatcatchers and Orchard Orioles. August is another good wader month, migrant passerines become more obvious and the first Ospreys, Bald Eagles and American Kestrels appear. September is an exciting month, especially for songbirds. On 16 September 2011, for example, after a cold front moved through and the prevailing wind swung to north-west, 27 species of wood-warblers were seen, including five Connecticuts, a rare east-coast bird. The following day, in a haul of 1,550 warblers there were more than 500 Northern Parulas. At other times, with slack southerlies, passerine migration may be all but invisible. As wader passage starts to tail off late in September, so birds of prey increase. Wildfowl numbers also build up and wave after wave of songbirds appear and disappear.

Red Knot

This medium-sized wader (*Calidris canuta*) undertakes one of the longest of all migrations twice a year: 15,000 kilometres in each direction. It breeds on high Arctic tundra in June and July. Many of the birds that breed in the far north of eastern Canada and Greenland use the intertidal zone of Delaware Bay as a stopover site. Adult females leave the nest sites in July, the males following some time later and the fledglings making the journey on their own. Some Red Knot winter in Florida, but others make the epic journey to the southern tip of South America. A few months later they repeat the trip in reverse, arriving in Delaware Bay in May, just as the horseshoe crabs there are coming ashore to spawn – their eggs are a welcome extra food source.

Early October witnesses the raptor peak, with impressive numbers of Merlins and Northern Harriers. Wildfowl, divers, seabirds and Great Blue Herons move along the coast. From mid-month it is the turn of thrushes, Yellow-rumped Warblers, sparrows and Red-winged Blackbirds to take centre stage. A Cape May regular, Mike Crewe, describes the never-to-be-forgotten experience of visible migration across Delaware Bay at the end of October: "I did not stop looking at birds – continually, without a break in their numbers – for well over an hour. Yes, many are common species, but it's the spectacle that is so amazing. Tight balls of Cedar Waxwings and European Starlings, compact lines of Red-winged Blackbirds (with the promise of a Rusty Blackbird tucked in between), spaced-out carpets of American Robins riding high above all the others and ragged, undecided packs of Blue Jays constantly testing the air waves. When the birds are high, you rely much more on your ears than your eyes ..." Other species are less dependable. Pine Siskins, for example, are irruptive rather than truly migratory. Their movements are dependant on the availability of food, rather than the weather, so they don't appear in big numbers every year.

Between 25 October and 15 November thousands of Northern Saw-whet Owls pass over at night; many of these are trapped by ringers in good 'owl years', along with smaller numbers of Long-eared Owls. Red-tailed and Red-shouldered Hawks are November birds, and short-distance passerine migrants, including kinglets and sparrows, stream through. Early November is a good time for finding rarities. And still the cavalcade is not over! December witnesses more coastal movements of divers and Gannets, and the very last passage of raptors and passerines.

Working the sites

Higbee Beach is the area's must-bird site. On its day, the mixture of woodland, grassland and scrub, with a few pools, is unbeatable because it provides cover for migrants that arrive during the night and food for them the following morning. In spring, slowly bird the trails and woodland edges. In autumn the dike (from where the morning flight counts are conducted) is the best place to start before sunrise and for a couple of hours afterwards. View the tree canopy from there. Later, work the trails and woodland margins. Scan the sky regularly for raptors, though if there are lots of passerines many accipiters will be hunting at treetop height.

A multi-levelled wooded raptor watchpoint in Cape May Point State Park is the best place to watch passing birds of prey, especially between mid-September and the end of November. On a good day (with north-westerlies) observers may witness the passage of 2,000–3,000 Sharp-shinned Hawks, dozens of Merlins and Peregrines, and hundreds of Northern Harriers. More

Northern Saw-whet Owl (*Aegolius acadicus*) is a partial migrant. Some remain in the coniferous and mixed forests of New England, Canada and the Appalachians. Others move south. This bird is an adult.

than 1,000 Ospreys have been seen in a single day. In late October and November there are good flights of Red-tailed and Cooper's Hawks and more Sharp-shinned Hawks and Northern Harriers. In autumn 2003 more than 45,000 birds of prey were counted. The passage of hundreds of Chimney Swifts and swallows can also be memorable.

Cape May Meadows, just east of the State Park, is a combination of grassland and freshwater pools next to mudflats. This is an ideal site for wildfowl, herons, egrets and waders from April onwards. Least Bitterns frequent the *Phragmites* around the pools, and Virginia Rails, Marsh Wrens and Common Yellowthroats also use the reedbeds. Between mid-September and the end of October raptor passage over the meadows can be spectacular, and the site is also good for sparrows and other passerines.

In May hundreds or thousands of Red Knot roost near the tip of Stone Harbor Point, 20 kilometres north-east of Cape May, having feasted on horseshoe crab eggs in Delaware Bay by day. The Stone Harbor roost also attracts Willet, Grey Plover, Sanderling, Dunlin and Ruddy Turnstone.

Avalon and Heislerville

The Avalon seawatch point juts out into the Atlantic at the north end of Avalon, which is 6 kilometres up the coast from Stone Harbor. It can be productive at any time of year, but autumn is best. Dozens of Arctic Skuas are seen annually, along with hundreds of thousands of Double-crested Cormorants, tens of thousands of Black and Surf Scoters, Red-throated Divers, Northern Gannets, gulls and terns. Scoters, divers and Northern Gannets stream past in late October and November. There is a chance of something rarer, a Pomarine Skua, for example. Passerines often 'coast'. For Delaware Bay waders, Heislerville Wildlife Management Area is the place to go. It is about 40 kilometres north of Cape May and faces Delaware Bay. Visit is when the tide is up in mid-May when there may be 30,000 Grey and Semipalmated Plovers, Greater and Lesser Yellowlegs, Short-billed Dowitcher, Least, Semipalmated and White-rumped Sandpipers, and Red Knot. Heislerville is also a regular site for Curlew Sandpiper, a rarity from the Old World, in May.

Scarce birds and rarities

Cape May's catalogue of rarities includes southern overshoots, astonishing seabirds and vagrants from the Pacific and Western Palearctic. In the first category come Swallow-tailed Kite and White-winged Dove, which are now almost annual in May, and Vermilion Flycatcher, a real 'mega'. Yellow-nosed Albatross has been seen in May, and Black-capped Petrel, Audubon's Shearwater and Band-rumped Storm-petrel in July. In autumn Ash-throated Flycatcher and Cave Swallow are pretty much annual but Mountain Bluebird (November 1988), MacGillivray's Warbler (November 1997) and Townsend's Solitaire (October 2012) are unlikely to be repeated any time soon – though at Cape May you would not bet against!

Blue-headed Vireo (*Vireo solitarius*) is a short- or medium-distance migrant, wintering in the southern United States and Mexico. It breeds in mixed woodland. It is one of the first passerine migrants through Cape May in spring and one of the last to leave in autumn.

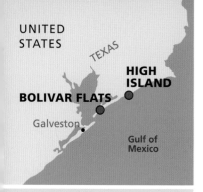

UNITED
STATES

TEXAS

HIGH
ISLAND

BOLIVAR FLATS

Galveston

Gulf of
Mexico

- **LOCATION**
  Gulf coast of Upper Texas,
  United States

- **FLYWAY**
  Central (and Mississippi)
  Americas

- **SPRING**
  Raptors, waders and
  passerines – especially
  thrushes, vireos, wood-
  warblers, grosbeaks and
  orioles – between early
  March and mid-May. Mid- to
  late April usually has the
  greatest variety.

- **AUTUMN**
  The first shorebirds return
  in late June. Wood-warblers
  are on the move by August.
  In September, more wood-
  warblers, sparrows, raptors,
  shorebirds and the first
  wildfowl appear.

- **KEY SITES**
  Smith Oaks and Boy
  Scouts Woods sanctuaries,
  Bolivar Flats, Anahuac.

- **CONSERVATION
  THREATS**
  The main sites are
  protected.

Upper Texas Coast, USA

High Island can often surprise even veteran migration birders. An early morning walk around one of its wooded areas in spring may produce next to no birds. Then all of a sudden, in early afternoon, the canopy may fill with bright, calling wood-warblers, tanagers and orioles. The latest shipment from Mexico has arrived!

The attraction of High Island for an exhausted passerine is undeniable. Imagine yourself as a vireo or wood-warbler that left Mexico's Yucatan Peninsula at dusk one day in mid-April en route for breeding grounds in the United States. Having flown 18 hours through the night and the following morning, your mental tachometer has clocked up more than 1,000 kilometres. You desperately need food and drink, and somewhere to bathe. The good news is that the Texas coastline is visible in the distance, but the bad news is that it seems featureless, bereft of cover. More serious, the wind has switched to a north-westerly during the morning, making flight more exhausting. However, on a slightly raised area just beyond the coast is a tract of woodland. A slight change of course will take you there …

High Island is a gently raised area sitting on a salt dome that rises a less-than-dramatic 10 metres above the plain but supports the area's best stands of woodland. It is an 'island' in the flat, treeless Texas coastal plain. That accident of geography, and the fact that generations of birders and conservationists have studied and managed the site, has rewarded it with one of the best track records of any migration hotspot. It is possible to see well over 100 species before lunch on a good spring morning.

On 27 April 2008, during a 'big sit', 136 species were logged just from the yard of the information centre!

Northern Parulas (*Parula americana*) first arrive on the Upper Texas Coast in late March on their way to breed in moist woodland as far north as mid-Ontario.

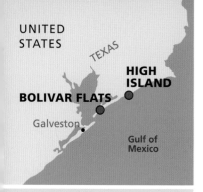

The birds

If weather conditions are good, most spring migrants will continue flying
after they hit the coast, aiming to find cover later in the day. If a fast-moving
cold front brings blustery northerly or north-westerly winds and showers,
a proportion of north-migrating birds will pitch down at the coast. This
produces a fall, which may involve a few score birds – or thousands, depending
on how many are in transit at the time. This explains two phenomena about
High Island. Typically, even during the peak of the spring migration, which
runs from mid-March to mid-May, several 'quiet' days may separate each
obvious arrival of wood-warblers, vireos, tanagers and orioles. Also, the length
of the flight from the Yucatan explains why High Island's woods can be very
quiet in the morning but become a scene of frantic bird (and birder!) activity
in the afternoon. Bird migration does not follow hard-and-fast rules, however.
Veterans of the upper Texas coast will admit that sometimes they forecast a
fall that never materialises. At other times the woods are jumping with birds
at dawn. That's the magic of birding!

**A trail at Smith Oaks
Sanctuary, High Island.
If an early session is
quiet here, don't give
up on the place. It is
worth returning in late
morning – sometimes
migrants are 'new in'
after 11 a.m.**

The seasons

Mid-March to mid-May and late September to mid-October are the most exciting periods, though with a broad range of wintering and breeding birds, the area could never be described as dull. At the beginning of March, wintering birds are moving out of the area and northbound migrants start to appear. The first include Black-and-white and Yellow-throated Warblers, Northern Parulas and – at Bolivar Flats – Stilt and Pectoral Sandpipers. Louisiana Waterthrushes, Prothonotary, Hooded and Kentucky Warblers, and Baltimore Orioles appear before the month's end.

By the second week of April daily logs will have registered 20 or more wood-warbler species. For example, the first really big day in 2009 was 10 April (24 species of wood-warblers), then in 2010 it was 18th (29 species), in 2011 it was 12th (24 species) and in 2012 it was 6th (21 species). At this time a subtle change in weather can deliver the goods, though this is far from guaranteed. For example, on 24 April 2010 the morning was quiet. Any birders who left would have cursed themselves, though, because in early afternoon the invasion started: wave after wave of Grey Catbirds, Rose-breasted Grosbeaks, Western Kingbirds, Summer and Scarlet Tanagers, Orchard and Baltimore Orioles (and one Scott's Oriole), vireos (including Cassin's) and wood-warblers appeared in the oaks. A Fork-tailed Flycatcher was in the mix as well. When the dust had settled, the wood-warbler count topped out at 32 species. Two years later, on 16 April, one keen counter estimated 600 Tennessee and dozens of Black-throated Green Warblers in Smith Oaks alone.

The expected and the unexpected

One High Island speciality is Swainson's Warbler; several of these skulkers are noted every year. Then there are the birds of prey, including Swallow-tailed and Mississippi Kites, which track along the Caribbean slope of Mexico rather than crossing the Gulf of Mexico. Sometimes more than 100 of the latter pass in a single afternoon in April, though Swallow-tailed are much scarcer. The frenzy calms in mid-May, but late-moving Black-billed Cuckoos, flycatchers, and Chestnut-sided, Bay-breasted and Magnolia Warblers can still be found, and Bolivar Flats is good for Hudsonian Godwits and White-rumped Sandpipers.

In August, Mississippi Kites pass through on their way from breeding grounds in the southern United States to Veracruz and South America. Their passage marks the start of autumn's raptor migration, which will see Broad-winged, Red-tailed, Swainson's and Red-shouldered Hawks and American Kestrels move through in numbers in September and October. Scarcer, but regular, are Peregrine, Merlin and Swallow-tailed Kite. Species such as Upland and Baird's Sandpipers appear in the area, as do the first returning passerines. The returning wood-warblers won't now be in their spring finest and will present far more identification conundrums. As October turns to November, expect wildfowl, the calling of Sandhill Cranes and the arrival of wintering landbirds such as Hermit Thrushes and Palm Warblers.

Buff-breasted Sandpiper (*Tryngites subruficollis*) uses different migration strategies in spring and autumn. In spring it can be seen on fields on the Upper Texas Coast before moving through the interior of North America to the high Arctic where it breeds to 80° north. In autumn most birds migrate further east.

Smith Oaks and Boy Scouts Woods

The two main sites are managed by Houston Audubon Society (HAS) and are in the small community of High Island. Smith Oaks Sanctuary is the *piece de resistance*. Its stands of oak are crucial 'migrant magnets' but there are also ponds, reedbeds and a small area of prairie. A good plan is to work this site for a couple of hours after dawn. If there are birds, you probably won't want to leave. If it is quiet, check out Boy Scouts Wood (almost adjacent) or Bolivar Flats (for waders) and return in early afternoon. When migrants are fresh in, Boy Scouts Woods Sanctuary is amazing. Its boardwalks meander through oaks and hackberries where birders can easily become afflicted by 'warbler neck'. There is even a grandstand for birders to sit on while watching birds come to bathe and drink at Purkey's Pond. In mid-April it is possible to see 15 or more species of wood-warbler without leaving your seat.

If things are quiet in the sanctuaries, there is no shortage of other places to work. For waders, head west along the coast for a few kilometres to Bolivar Flats, another HAS reserve. This invertebrate-rich mix of saltmarsh and intertidal mud and sand provides food for thousands of migrant, wintering or breeding wildfowl, herons, waders, gulls and terns. The variety of passage waders is particularly impressive. This is one of the most important sites for waders in the United States. The list of possibilities is virtually a roll-call of North America's breeding species, including Wilson's, Piping, Snowy and

Cerulean Warbler

Small numbers of this *Setophaga* pass through High Island every April. With its vivid sky-blue upperparts, a spring male is guaranteed to set any birder's pulse racing. Sadly, breeding numbers fell by a staggering 83 per cent in the last four decades of the 20th century, a decline that justifies its status as Vulnerable. Ceruleans breed in mature deciduous forests (usually near swamps) in the eastern United States and parts of eastern Canada. Their main migration routes to winter quarters in northern South America are through the south-eastern states of the United States to Cuba, Jamaica and South America; and through the Caribbean slope of Mexico and Central America. Mountaintop mining (in West Virginia, Virginia, Tennessee and Kentucky) and general forest fragmentation degrade breeding habitat, while the conversion of shade coffee plantations (their favoured winter habitat) to sun coffee in Colombia and Venezuela is believed to have hit populations as well.

Semipalmated Plovers, Red Knot, and Least, Western, Semipalmated and Stilt Sandpipers in the intertidal zone and wetland shallows. Fields can be good for American Golden Plover and Upland and Buff-breasted Sandpipers. Herons, pelicans, skimmers, gulls and (sometimes) nine species of terns can be seen. Check out the roosts when the tide is up. In the opposite direction, only about 15 kilometres away, is Anahuac National Wildlife Refuge. Apart from its great assortment of breeding herons and rails, any areas of cover can host flycatchers, wood-warblers, grosbeaks, sparrows and orioles in spring. Although far from 'classic' wood-warbler habitat, 20 species have been seen in a day, so check any stands of willows. If water levels in the pools are low in autumn, there should be waders.

A male Baltimore Oriole (*Icterus galbula*). Some migrate along the Gulf coast of Mexico, while others cross the Gulf, where many make landfall at High Island in early April. This species does not breed locally but continues northwards.

Recent rarities

A Red-necked Stint was at Bolivar Flats in June 2011, and High Island scored with a Black-whiskered Vireo in April 2009, a Sulphur-bellied Flycatcher in April 2011 and the first U.S. record of Double-toothed Kite the next month.

TEXAS

Gulf of
Mexico

UNITED
STATES

SOUTH PADRE
ISLAND

LAGUNA
ATASCOSA

Rio Grande • Brownsville

MEXICO

South Padre Island, USA

Ideally placed by the mouth of the Rio Grande, South Padre Island and Laguna Atascosa are decidedly more Mex than Tex. Hundreds of species of migrants pass through on their way around the western end of the Gulf of Mexico.

- **LOCATION**
  Gulf coast of Lower Texas, United States

- **FLYWAY**
  Central (and Mississippi) Americas

- **SPRING**
  Purple Martins are one of the first species to move, from late January. They are followed by wildfowl, raptors, waders, hummingbirds, tyrant flycatchers, wood-warblers, sparrows and many more. The last northbound birds are seen in early June.

- **AUTUMN**
  The first returning Orchard Orioles are often seen in July. Offshore, seabirds can be interesting in August, when wader passage steps up a gear. Then wildfowl, passerines, cranes and raptors extend the season into November.

- **KEY SITES**
  World Birding Center and Sheepshead Lots, South Padre Island; Laguna Atascosa National Wildfowl Refuge.

- **THREATS**
  Sites are protected, though disturbance can be a problem on parts of South Padre Island.

South Padre Island is a very long, narrow barrier island that runs south-north, with the Gulf of Mexico to the east and the shallow, sheltered, hypersaline Laguna Madre shimmering between it and the Texas mainland. Parts of South Padre are like a cross between a holiday resort and a subtropical Scilly, with beachfront hotels and bars along Gulf Boulevard, and surfer dudes enjoying the Gulf waters, but the atmosphere in spring and autumn is pregnant with the prospect of rare birds. South Padre's geography puts it in pole position to attract migrants and there are some excellent patches of habitat. The contrast is dramatic between the low number of resident species and 200 or so species of non-breeding migrants, many of them scarce or rare species.

Along the west coast of South Padre Island are intertidal mudflats, areas of saltmarsh and dune meadows. The east coast is mainly sandy beach. As with any place where migrants make first landfall or reach the daunting barrier of the ocean, any patch of cover, or pool for bathing or drinking, can and does attract birds when conditions are right.

Parts of the island are included within the huge Laguna Atascosa National Wildfowl Refuge, most of which is on the west side of Laguna Madre. The Laguna Atascosa reserve is huge: 36,000 hectares of thorn scrub, prairie, lakes and seasonal pools. It was established in 1929 "for use as an inviolate sanctuary… for migratory birds" and has been doing just that ever since. The Queen Isabella Causeway connects the southern end of South Padre with the town of Port Isabel and the bulk of the refuge.

A MacGillivray's Warbler (*Geothlypis tolmiei*) rests after making landfall. This species does not breed in Texas, which is east of its usual migration route between Mexico and the western United States.

Tex-Mex avifauna

Well over 400 species of birds have been seen at Laguna Atascosa, and its checklist has a distinctly Mexican feel, hardly surprising as it is just a few kilometres north of the Rio Grande with the Mexican state of Tamaulipas beyond. South Padre–Laguna Atascosa is at the very northern extremity of the breeding range of some Neotropical species, such as Northern Beardless-Tyrannulet, Brown-crested Flycatcher and Botteri's Sparrow. In spring it receives overshoots of others, like Sulphur-bellied Flycatcher, Rose-throated Becard and Broad-tailed Hummingbird. And it is also on one of the main conduits for passerines and raptors that winter in Central and South America and breed in the United States and Canada. South Padre Island also picks up spring migrants that have launched from the Yucatan peninsula, Mexico, and been drifted west by onshore winds. The area's pools and intertidal zones attract tens of thousands of waders, gulls and terns tracking to and from breeding colonies along the coast, and – although not one of the world's great seabird oceans – the Gulf of Mexico does have some interesting pelagic species, especially during the summer. These are sometimes pushed close to South Padre Island.

The seasons

The only times when migration is not in evidence are short periods in midwinter and midsummer. Purple Martins start to pass north in late January and Couch's Kingbird (a short-distant migrant) moves the following month. By the start of March migrant House Wrens, sparrows and early wood-warblers, especially Black-throated Green, will be evident. In March 2012 a Golden-cheeked Warbler, which has a very limited breeding range in the Texas hill country, was found at the Sheepshead Lots. Later in the month Broad-winged and Swainson's Hawks and a few Mississippi Kites keep the parade going, and Anhingas may join the kettles of raptors. Lesser Nighthawks, Palm Warblers, Louisiana Waterthrushes, Northern Parulas and Great Crested Flycatchers are also seen.

Eastern Kingbird (*Tyrannus tyrannus*) is a summer visitor to the United States, wintering as far south as northern Chile. A diurnal migrant, it often moves in small, loose groups. This flycatcher does not breed in southern Texas but it may pass through South Padre from late March.

April sees a big push of birds of prey and the peak passage of waders, nighthawks, Yellow-billed Cuckoos, hummingbirds (migrant Ruby-throated and Black-chinned are expected, and Buff-bellied is resident), swallows and most warblers. As with High Island, further up the Texas coast, weather has a big effect on which migrants are actually observed. Cold fronts with rain and north-westerly winds often produce falls at South Padre Island, particularly if they follow easterly breezes. Bear in mind that for birds that have flown over the Gulf from the Yucatan Peninsula, arrival will not be before 11 a.m. or midday. If conditions are particularly bad they may not make landfall until 5 p.m. On the other hand, benign southerlies will allow northbound birds to fly straight over.

Throughout the spring, Laguna Madre and the Laguna Atascosa *resacas* (shallow lakes) empty of wildfowl, while waders, gulls and terns stream north up the coast. Passage slows in May, although early in the month is the best time to look for late-season migrants such as Brown-crested and Willow Flycatchers and Magnolia, Blackburnian and Bay-breasted Warblers. The tardiest migrants may put in an appearance even in early June. Just a few weeks later, in early July, Orchard Orioles, *Empidonax* flycatchers, Purple Martins and other early returnees begin to filter south.

Summer and autumn

A few pelagic trips run out of South Padre in midsummer. A successful trip aboard the *Osprey* at the end of August 2011 produced Sooty, Bridled, Black, Least and Common Terns, five Band-rumped Storm-petrels and a Leach's Storm-petrel, five Audubon's Shearwaters and several Masked Boobies. A few passerines included Yellow and Wilson's Warblers and an Olive-sided Flycatcher, way offshore.

Autumn migration is prolonged: wave after wave of waders, birds of prey and songbirds entertain birders through to late November. Red Knot and Piping Plover return in September, with the former just passing through and the latter staying on the flats near the Convention Centre until April. South Padre is also good for wood-warblers, tanagers and orioles in September; for returning wildfowl and sparrows in October; and for rarities in October and November. Large numbers of passerine migrants may still show up on the island in mid-November. Sandhill Cranes return to Laguna Atascosa from their breeding grounds in Canada and the north-west United States, and hundreds of thousands of ducks appear on Laguna Madre, including 80 per cent of North America's Redhead population. Greater White-fronted and Snow Geese, Gadwall, American Wigeon, Blue-winged Teal, Northern Shoveler, Northern Pintail, Green-winged Teal, Canvasback and Ruddy Duck are also either common or abundant there and in the Laguna Atascosa reserve. Ospreys return in good numbers, and Northern Harriers, Red-tailed Hawks and American Kestrels come back to hunt over Laguna Atascosa, joining the reintroduced local breeder, Aplomado Falcon. Countless northern-breeding sparrows and buntings use the reserve as their winter domicile.

A male Black-chinned Hummingbird (*Archilochus alexandri*). Like so many other species that pass through South Padre Island, it does not stay to breed. It may turn up between March and May and in August or September. This hummer is a medium-distance migrant, wintering in Mexico and breeding in the United States west of the Rockies.

Working the sites

There are two stand-out locations on South Padre Island. The first is the area around the South Padre Island Birding and Nature Center and the Convention Center, a few kilometers north of the island's southern tip. The Birding and Nature Center is one of the nine World Birding Center locations in the Lower Rio Grande Valley. It has dune meadows, saltmarsh, freshwater pools and shrub thickets. About 1.5 kilometres of interconnected boardwalks twist through the marsh, providing great opportunities for watching herons, rails and waders. Where well-watched rarities are taken by Sparrowhawks on Scilly, on South Padre the offending predator may well be an alligator! The 20-hectare site has been designed for wildlife and has the greatest variety of species on the island. A walk around this small area in spring or autumn can produce more than 80 species. The pools attract anything from bathing sparrows to Louisiana Waterthrushes. Just to the north, the trees and shrubs around the Convention Center are the nearest the island has to woodland and so exert a magnetic pull on arboreal species. Several species of hummingbirds feed at flowers; and the butterfly garden is a good place to seek out passerine migrants. The very long list of migrants seen around here in autumn 2011 included such very different species as Long-eared Owl, Northern Flicker, Anna's Hummingbird, Black Phoebe, Hammond's Flycatcher, Bewick's Wren, Western Tanager and Fox Sparrow.

The other key location on the island is on Sheepshead Street, about 4 kilometres to the south. The Sheepshead Lots (shrubby copses) are small but have 'drips' for bathing and drinking and fruit for visiting orioles. The site has an excellent reputation for wood-warblers but just about anything can turn up. In late October and early November 2011 it attracted Broad-tailed Hummingbird, Blue Bunting, Green-tailed Towhee and Pine Siskin. A couple of other places to check are Andy Bowie Park, on the opposite side of State Park Road from the Convention Center, and Isla Blanca Park, which is 2 kilometres south of the Sheepshead Lots. The park's jetty looks out over the Gulf of Mexico and can be reasonable for seawatching: Masked Booby, skuas, gulls and terns are seen from here. In winter it is a good place to watch divers and scoters.

Just across Laguna Padre on the mainland, the Laguna Atascosa reserve has several breeding specialities (Plain Chachalaca, Greater

Roadrunner, Verdin and Long-billed Thrasher, to name but a few) in addition to its passage migrants and hordes of winter wildfowl. More than 30 species of wildfowl are known to have used the reserve, including Texas rarities Surf Scoter, Masked Duck and Eurasian Wigeon. Grey and Zone-tailed Hawks and Common Black-hawk – primarily Mexican species – have all been recorded here in autumn while, from the other direction, Rough-legged Buzzard has been seen in winter and Northern Goshawk in spring. Other well-represented groups of migrants are tyrant flycatchers (24 species, including Olive-sided and Ash-throated, and Black Phoebe), wood-warblers (45 species, although only two breed) and sparrows (31 species).

Bayside Drive is a 25-kilometre, one-way loop through thorn forest and coastal prairie and alongside Laguna Madre. At the right time of year the shoreline has Grey, Snowy, Wilson's and Semipalmated Plovers, Western and Least Sandpipers, both dowitchers, Ospreys and many more species.

Lakeside Drive is much shorter, leading from the visitor centre to the shallow lake that gives the reserve its name, where the Osprey Outlook hide gives views over the assembled wildfowl – and American Alligators! Several walking trails also lead from the visitor center.

Sandhill Cranes (*Grus canadensis*) arriving at Laguna Atascosa – an important wintering site – in autumn. These cranes breed as far north as the north coast of Alaska.

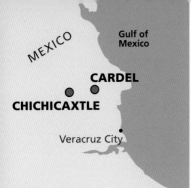

MEXICO

Gulf of
Mexico

CARDEL

CHICHICAXTLE

Veracruz City

- **LOCATION**

  *Close to the coast of the
  Gulf of Mexico in Veracruz
  region, Mexico*

- **FLYWAY**

  *Converged Pacific, Central,
  Mississippi and Atlantic
  American flyways*

- **SPRING**

  *Mostly Broad-winged and
  Swainson's Hawks and
  Mississippi Kite.*

- **AUTUMN**

  *Several million raptors,
  mostly Broad-winged
  and Swainson's Hawks,
  Mississippi Kite and Turkey
  Vulture. Also, Swallow-tailed
  Kite, Osprey, Northern
  Harrier, Sharp-shinned,
  Cooper's and Zone-tailed
  Hawks, American Kestrel
  and Peregrine; American
  White Pelican, Wood
  Stork, Anhinga, Roseate
  Spoonbill; White-winged
  Dove, Eastern Kingbird,
  Scissor-tailed Flycatcher
  and other passerines.*

- **KEY SITES**

  *Cardel, Chichicaxtle,
  Chavarillo and Tlacotalpan
  for raptors; Juan Angel
  beach and La Mancha are
  good for near-passerine
  and passerine migrants.*

- **THREATS**

  *Main threats are to the
  birds' breeding and
  wintering areas.*

Veracruz, Mexico

*The Veracruz 'River of Raptors' is one of the most
dramatic and inspiring sights in the natural world.
It flows between August and November and carries the
world's biggest concentration of migrating birds of prey.*

Unlike Spain's Tarifa or Italy's Messina, the Veracruz bottleneck is not a 'jumping-off' point for birds looking for the narrowest sea crossing. It has more in common with Chumphon in Thailand and Batumi in Georgia, where birds seek a passage between mountains and sea. Each autumn about 25 species of migratory raptors pass along the coastal strip near the city of Veracruz, especially around the towns of Cardel and Chichicaxtle.

Virtually the entire world populations of Mississippi Kite, Broad-winged Hawk and Swainson's Hawk are involved. The first of these is en route from its disjointed breeding range in the southern United States to wintering quarters in South America. Broad-winged Hawk breeds across much of eastern North America and spends the winter in northern South America, while its western North American counterpart, Swainson's Hawk, is one of the world's longest-migrating birds of prey, wintering on the pampas of Argentina and southern Brazil and thus making a round-trip of 20,000 kilometres.

Enormous numbers of Turkey Vultures are also involved in the River of Raptors, along with smaller counts of Osprey, Swallow-tailed and Hook-billed Kites, Northern Harrier, Sharp-shinned, Cooper's, Grey, Zone-tailed, Red-tailed and Red-shouldered Hawks, American Kestrel, Merlin and Peregrine. They are joined in the thermalling mixture by American White Pelicans, Wood Storks and Anhingas.

Between mountains and sea

Since the early 1990s Pronatura Veracruz, Hawk Mountain Sanctuary and HawkWatch International have coordinated counts in and around Cardel and Chichicaxtle. The figures make astonishing reading: 2 million Broad-winged Hawks, 1.5 million Turkey Vultures, 1 million Swainson's Hawks and 200,000 Mississippi Kites. Between 5 million and 6 million raptors use the route, along with an estimated 10 million near-passerine and passerine diurnal migrants.

Every autumn tens of millions of migrants, including birds of prey, move south through the eastern Mexican lowlands. As the lowlands narrow, squeezed between the Sierra Madre Oriental and the Gulf of Mexico, so the broad-winged raptors bunch up since, for them, travel by thermal is the most efficient way of covering long distances. Large broad-winged birds – *Buteo*

Red-tailed Hawks (*Buteo jamaicensis*) move through Cardel in autumn, though in smaller numbers than other species of *Buteo*.

species, vultures, pelicans and storks – rely on thermals of rising air far more than *Accipiter* species and falcons, and are better equipped to take advantage of them. Thermals develop best over low ground, whereas they are erratic over mountainous country and do not get going at all over water. North of the historic port of Veracruz, in Veracruz state, the mountains almost reach the coast, creating a bottleneck for soaring birds just 20 kilometres wide. Cardel and Chichicaxtle are just south of the bottleneck.

When the ground begins to heat up, typically from about 10:00 hrs in the morning, thermals of rising warm air are generated and a few birds find themselves being lifted. As they spiral higher, others see what is happening and fly to the base of the thermal to take full advantage. Within minutes, a 'kettle' or vortex of upward-spiralling birds is formed, sometimes 1,000-strong. At the top, the birds glide out towards the next one. Rob Bierregaard describes what happens when birds leave a large kettle: "Their wings are set and tucked in as they're on a long, controlled glide to the south. If they're on a really long glide they will start to stratify – the heavier Turkey Vultures 'sink' to the bottom of the stream, the lighter Broad-wings 'float' to the top, with the Swainson's in the middle."

On the best days, particularly after a spell of poor weather when large numbers of hawks and vultures have been grounded, there is a rush to get going again. One such day was 17 October 2003, when out of a total of 1,512,973 raptors there were 782,653 Swainson's Hawks.

The 'river' in full flood

Mississippi Kites dominate proceedings from the third week of August, peaking in the first week of September. Their passage is very concentrated; up to 96,000 have been seen in a day. Swallow-tailed Kites are most likely to be seen in late August, albeit in much smaller numbers. Turkey Vultures are prominent whenever birds of prey are on the move, throughout the autumn. Many American White Pelicans and Anhingas and fewer Wood Storks are likely, with White and White-faced Ibis possible. Passerine diurnal migrants in early autumn include Eastern Kingbird, Blue-grey Gnatcatcher, Dickcissel and Baltimore Oriole. Up to 1,000 of the first species have been counted in a day in mid-September.

From mid-September Broad-winged Hawks begin to appear in large, very large or phenomenal numbers. On 26 September 2002, 626,692 were logged, including 372,000 between midday and 1:00 hrs – 100 birds per second! Another stand-out day was 28 September 2010, when a combined tally made by counters at Cardel and Chichicaxtle (10 kilometres to the west) produced 453,260 Broad-winged Hawks out of a raptor total of 473,944. In 2012 the peak day for Broad-wings was 2 October, when 286,131 Broad-winged Hawks and 26,175 Turkey Vultures were counted.

Broad-winged Hawk is still the major player in early October, but as its numbers decline, those of Swainson's Hawk climb, peaking in the middle of the month. A classic date for this species was the previously mentioned 17 October 2003, when more than three-quarters of a million Swainson's were counted. More than 10,000 American White Pelicans were also noted that day, and other guaranteed non-raptors on the move in October include Anhinga, Wood Stork, White-faced Ibis, White-winged Dove and Scissor-tailed Flycatcher. Numbers dwindle as November progresses, but plenty of Turkey Vultures continue to kettle their way south, while smaller numbers of Sharp-shinned and Cooper's Hawks, American Kestrels, Merlins and Peregrines take a more direct flightpath.

A big kettle of raptors is an inspiring sight. Kettles can form quickly, and on a hot day, when warmed air is rising rapidly, the birds at the top can be out of sight within a few minutes. In early October Broad-winged Hawks (*Buteo platypterus*) dominate proceedings, but Swainson's Hawks (*Buteo swainsoni*) take over in the second half of the month.

Broader front migration in spring

Spring raptor passage occurs between the first week of March and the second week of May. It takes place on a broader front, so is not concentrated at any one site, but Tlacotalpan (60 kilometres south-east of Veracruz city) has proven to be good. Contrasts with autumn are marked. Twenty raptor species have been noted passing over Tlacotalpan, the most prominent again being Broad-winged and Swainson's Hawks and Mississippi Kites. Turkey Vultures pass north in only relatively small numbers. Mississippi Kites average about 20,000 each spring; they start to come through in the last week of March, with a peak in the last week of April, thereafter tailing off rapidly. An average of 80,000 Broad-winged Hawks head north between the first week of March and the second week of May, with the peak in the first week of April. Swainson's move in the same time-frame, averaging fewer than 3,500 but with a remarkably consistent peak date between 16–18 April.

Raptor watchpoints

Cardel is 30 kilometres north-west of Veracruz city. The spectacle can be viewed from anywhere around the town, but to watch the expert raptor counters at work and support the efforts of Pronatura Veracruz, six floors up on the roof of the Hotel Bienvenidos is the place to be. Since big movements are dependent on thermals, not much happens before 10:00 hrs. Later in the day, an easterly breeze may push in from the Gulf of Mexico, drifting the soaring birds further inland. Pronatura's Migratory Bird Observatory at Chichicaxtle, about 10 kilometres to the west, is then the best alternative. Tlacotalpan is still good for raptors in spring, and counts are also conducted at Chavarillo, 15 kilometres south-east of Jalapa city, with spring totals averaging 300,000.

The raptor watchpoints of Veracruz monitor non-raptors as well, including Roseate Spoonbills (*Platalea ajaja*). Some migrate along the Gulf of Mexico coastal plain in spring and autumn.

SOBERANIA NP

Panama City ● ○ UPPER BAY

Pacific Ocean

- **LOCATION**
  Pacific coast of Panama

- **FLYWAY**
  Atlantic and Central
  Americas

- **SPRING**
  Waders peak in Upper Bay
  in February and March,
  while raptor passage is
  concentrated in March
  and the first three weeks
  of April.

- **AUTUMN**
  Return wader migration
  develops from a trickle
  in late July to a flood in
  September and October;
  Mississippi Kites often
  herald the start of the
  raptor season in late
  September, with Broad-
  winged and Swainson's
  Hawks and Turkey Vultures
  peaking in October and
  November.

- **KEY SITES**
  Cerro Ancon, Semaphore
  Hill Canopy Tower,
  Rainforest Discovery
  Center, Upper Bay of
  Panama.

- **THREATS**
  Urbanisation of the Upper
  Bay of Panama.

Panama City, Panama

For sheer numbers of migrating raptors, it is hard to beat the hills around Panama City. In recent years it has become the norm for 2 million to pass over the city, or at least close by, in September and October. This is one of the great birding spectacles.

The key players in this avian drama are Turkey Vultures, Broad-winged and Swainson's Hawks and, to a lesser extent, Mississippi Kites. While Turkey Vultures are sedentary over most of their range, those that breed in Canada and much of the interior of the United States are frozen out in winter and have to move south, many to northern South America. Broad-winged Hawks abandon the eastern United States and Canada for the same reason. And Swainson's Hawks take their migration even further: they leave western North America for the pampas of Argentina. Virtually all the Swainson's, most of the Broad-winged Hawks and a good proportion of the Turkey Vultures have to pass through the narrow S-shaped isthmus of Panama to reach their destinations. Unwilling to cross the Pacific Ocean to the south, this 'river of raptors' is extraordinarily concentrated – a classic raptor bottleneck – as in eastern Veracruz, Mexico, 2,000 kilometres to the north-west.

Spectacular numbers

In 2010 and 2011 an average of 1.8 million vultures, kites, hawks and falcons passed over Panama City's own watchpoint of Cerro Ancon between the beginning of October and the end of the third week of November. On some days the passage was almost unbelievable. For example, on 27 October 2011, 548,740 Turkey Vultures and 248,371 Swainson's Hawks were counted; on 8 October 2010, the Broad-winged Hawk count was 407,516; and 10,804 Mississippi Kites were logged on 4 October 2010. The average (2010–2011) autumn tallies for the four most common species at Cerro Ancon were: 935,000 Turkey Vultures, 359,000 Broad-winged Hawks, 226,000 Swainson's Hawks and 11,500 Mississippi Kites.

After the 'big four' of Turkey Vulture, Broad-winged Hawk, Swainson's Hawk and Mississippi Kite, other raptors by order of frequency are Swallow-tailed Kite, Peregrine Falcon, Osprey, Northern Harrier, Merlin, Cooper's Hawk, Zone-tailed Hawk and Red-tailed Hawk, though the last few do not pass in big numbers.

Spring migration is more diffuse but nevertheless still impressive. Turkey Vultures, Broad-winged Hawks and Swainson's Hawks are still the major players, but Mississippi Kite numbers are much smaller than in autumn.

A vulture's-eye view of Soberania National Park, north of Panama City, with the Centennial Bridge over the Panama Canal in the background. The Canopy Tower's observation deck is the ideal place for eyeballing migrant raptors, and the forest is home to hundreds of resident species and – during the winter months – North American breeders such as wood-warblers and vireos.

Wader staging post

Just east of Panama City, the Upper Bay is one of the most important areas for migrant waders in the Americas. This invertebrate-rich staging post is used by 1 million to 2 million waders in autumn, including half the world's female Western Sandpipers (and a third of that species' total population). The Upper Bay has rocky, sandy and muddy intertidal areas as well as adjacent grassland and wetland habitats. This mix appeals to a broad range of species. As well as Western Sandpiper, waders range from Surfbird and Ruddy Turnstone in rocky areas to Sanderling, Red Knot, Semipalmated, Least, White-rumped, Pectoral, Baird's, Buff-breasted and Upland Sandpipers, Hudsonian and Marbled Godwits, Greater and Lesser Yellowlegs and more on the sand and mud. Spectacular day counts have included 282,000 Western and 47,000 Semipalmated Sandpipers; 31,000 Semipalmated Plover; 12,000 Short-billed Dowitcher; and 7,000 Whimbrel.

Spring and autumn

In February and March numbers of Western Sandpipers (and other waders) swell as the wintering population is supplemented by birds moving north from as far afield as Peru. After this time they decline, though about 10,000–12,000 non-breeding young birds remain in the bay all summer. Turkey Vultures move through from the start of March and they are joined in mid-month by Broad-winged and Swainson's Hawks and Ospreys. By the end of the third week of April the main raptor passage has passed.

Early autumn migrant waders appear in the bay in late July, build through August and peak in October, by which time there may be close to 300,000 Western Sandpipers alone in the zone east of Panama City. Numbers decline thereafter. While spring raptor passage is impressive it is autumn when it really shines. Of the commoner species, Mississippi Kites are typically the first to peak, at the start of October, followed by Broad-winged Hawks around the middle of the month. Swainson's Hawks reach their greatest concentration at the end of the month and Turkey Vultures increase in November. Not every day is good, even in peak season. One factor that causes a stop-start flow is the prevalence of tropical storms that hit Central America each autumn. When a storm strikes, raptors sit tight and stack up in large numbers. Visible migration grinds to a halt. After each storm passes, huge numbers of vultures and hawks take to the skies and this is often when the biggest counts are made.

Cerro Ancon, Canopy Tower and Rainforest Discovery Center

The three stand-out raptor watchpoints are Cerro Ancon, just south-west of Panama City; the Canopy Tower on Semaphore Hill, to the north-west of the city; and the Panama Rainforest Discovery Center, a little further out.

While many populations of Turkey Vulture (*Cathartes aura*) are sedentary, millions move out of the northern part of their range in Canada and the United States each autumn. More than 1 million of these pass through Panama on their way to northern South America.

Swallow-tailed Kites (*Elanoides forficatus*) pass in smaller numbers than Turkey Vultures and Broad-winged and Swainson's Hawks, but their passage is noteworthy nonetheless. For example, on 4 October 2010, 103 of this most graceful species were logged as they drifted over Panama City.

For raptor watching alone, the pick of these is probably Cerro Ancon. It is impossible to miss Cerro Ancon because a huge Panamanian flag flies from the 200-metre summit of this, the most prominent hill in Panama City. The top is accessed via a single-track road through some nice forest.

However, the other two locations offer an amazing all-round birding experience and should not be missed. For example, the forests of Soberania National Park, which surround the Canopy Tower, have a known avifauna of 603 species, including 32 wood-warblers that pass through or spend the winter here. And the tower, which stands on a 300-metre hill, offers great raptor-watching. It is possible to walk around its observation deck (the roof of a 'recycled' U.S. radar station, turned ecolodge) to get a 360-degree panorama of forest canopy and sky. The Canopy Tower is about halfway between the city centre and the third site, the Panama Rainforest Discovery Center. This is near Pipeline Road, close to Gamboa, north of Panama City. A metal 40-metre observation tower is ideal for viewing migrant birds of prey and, again, the surrounding forest is superb.

The Upper Bay

It is not possible to work the whole Upper Bay because of its extent. An estimated 80 per cent of its waders use a 30-kilometre stretch east of Panama City. After checking the tides (visit during an incoming tide), a good place to witness some of these shorebirds, especially in September or October, is from the shoreline drive in Costa del Este, just east of the causeway leading out of Panama City. This will give you a good idea of the scale of the wader concentration. High-tide roosts here are on various rock outcrops just offshore.

- **LOCATION**
  Southern tip of Sweden

- **FLYWAY**
  East Atlantic

- **SPRING**
  Wildfowl, waders, terns, chats and thrushes.

- **AUTUMN**
  Wildfowl; waders; skuas, Little Gull and terns; Short-eared and Long-eared Owls; swallows and martins; pipits and wagtails; chats and thrushes; warblers including Barred, Yellow-browed, Pallas's, Dusky and Radde's; Red-backed and Great Grey Shrikes; finches.

- **KEY SITES**
  Nabben, Storevang, Ljungen, South Flommen, Falsterbo Canal, Fyledalen, Krankesjön.

- **THREATS**
  Housing and leisure development.

Falsterbo, Sweden

It has been estimated that 500 million birds move through southern Sweden each autumn in advance of the harsh winter. A good proportion of them pass through Falsterbo, one of the world's top locations for autumn visible migration.

Falsterbo has many claims to fame. It was once an important Hanseatic League port. Its horse show is internationally renowned, as is its 18-hole golf course. And this, the south-western extremity of mainland Sweden, has the country's oldest lighthouse. In fact, its geography is the key to its most important claim to fame – as northern Europe's most exciting visible-migration site in autumn.

Every autumn a huge avian exodus takes place with birds streaming out of Sweden, Finland and north-west Russia before the bitter winter sets in. Most of them head south-west and millions of them are making their first serious journey. Many find themselves funnelled into an increasingly narrow peninsula at the very south-western tip of Sweden – the Falsterbo peninsula, which points a 10-kilometre finger towards Denmark. From Falsterbo they can take the shortest route across the open waters of the Skjutomrade to Denmark, just 24 kilometres away – and beyond. Hundreds of thousands can be seen doing just that. Of course, many more pass over unobserved at night, not all use the Falsterbo route and many of the diurnal travellers also go completely unnoticed.

The short route to Denmark

Nonetheless, Falsterbo at its best is one of the top 'vis-mig' places there is: raptors, pigeons, owls, hirundines and passerines often take the shortest route over the sea while wildfowl, waders, skuas, gulls and terns track the coast. Unlike most of the world's best migration hotspots, there is a long series (going back to 1973) of species-by-species counts for every diurnal migrant passing between 1 August and 20 November, during which period a team of counters is out every day.

This systematic counting has proved to be invaluable for tracking population trends and has produced some staggering figures. Between 1973 and 2011, some 63.5 million individual birds were logged in the autumn 'vis-mig' watches, quite apart from the consistent efforts of the ringing teams by the lighthouse. There is also something definitely quirky about birding on a popular golf course. Birders and golfers generally respect each other's different passions but there have been 'incidents' when stray balls have landed among the ranks of telescopes.

The timing of the heaviest passage of Rough-legged Buzzards (*Buteo lagopus*) is remarkably consistent, generally between 12–22 October. The years 2010 and 2011 were particularly good, with 1,991 and 2,380 noted, respectively. The best day ever was 13 October 2010, when 1,202 were counted.

Waders lead the way

Falsterbo is most definitely at its best in autumn. The season's action starts as early as late June with passage waders and Common Crossbills (some years are also good for Parrot Crossbills), then continues through July with more waders and Common Starlings. By August, pipits, wagtails, hirundines and wildfowl are trickling, then streaming, through. One of the first passerines to move in large numbers is Yellow Wagtail. Thousands of these roost at regular reedbed sites along the coast nearby before moving through Falsterbo in late August and early September. Tens of thousands are counted annually, though numbers have fallen markedly since the 1980s.

The first big raptor movements are at the end of the month, with hundreds of European Honey-buzzards, sometimes up to 1,500 passing through in a single day and averaging 7,000 a season. The mantle of most plentiful raptor passes to Common Buzzard and European Sparrowhawk as September progresses. The latter outnumbers any other raptor species, with daily totals in excess of 1,000 reasonably common in late September. Other regulars are Black and Red Kites, Western Marsh, Hen, Pallid and Montagu's Harriers, Rough-legged Buzzard, Lesser Spotted Eagle, Osprey, Common Kestrel, Merlin, Eurasian Hobby and Peregrine.

The sixteenth green at Falsterbo golf course, one of the best sites on the planet for watching the visible migration of passerines in autumn. The copse by the lighthouse (in the distance) is a superb spot to find nocturnal migrants.

When large numbers of diurnal-migrating passerines are moving low, it is commonplace to see European Sparrowhawks and the occasional Merlin hunting just a few metres above Falsterbo's hallowed greens and fairways. Even well into November raptor passage is still evident, Goshawk being a notably late mover.

And there's always the tantalising chance that a rare *Aquila* eagle may join the show. Greater Spotted has been almost annual since the start of the new millennium; there were two Lesser Spotted in October 2007 and October 2011; Steppe was annual from 2003–2009; and an Eastern Imperial Eagle was seen in mid-October 2007. Single Eleonora's Falcons were seen in September 2002 and September 2006. In total, 31 species of raptors have been seen, not on the same level as Eilat or Messina but not too shabby.

Record-breakers

By the beginning of October the sheer weight of numbers of raptors and passerines can be truly awe-inspiring, with up to half a million birds of many species on the move in a single day. At this time, huge mixed flocks of Chaffinch and Brambling often dominate proceedings and day-counts of 300,000 or more are not that rare. Multi-layered visible-migration is a feature. Stock Doves and Wood Pigeons may occupy the highest band, with corvids one level down, then flocks of Common Starlings and finches relatively low and tits sometimes flying at head height. The latter, especially Blue Tits, sometimes pass through in thousands. For example, in 2012 there was a record-breaking exodus of the species. More than a quarter of a million were counted during the last week of September and the first week of October, including an astonishing 87,500 on one day. Thousands of these were ringed.

No two years are the same. Spotted Nutcracker is prominent in irruption years, but virtually non-existent at other times. For example, 447 flew over on one October day in 1995. Turning the clock back even further, 260 Black Woodpeckers were noted on migration in the period 1973–1975. These days the species is scarcely annual. In 1994 there was a huge movement of Jays, with 5,000 seen in a day, gaining height, balking at the prospect of crossing the water, then trying again until they finally went for it. Falsterbo has the numbers but not the rarity track record of Öland, a long, thin island off Sweden's Baltic coast, about 300 kilometres north-east of Falsterbo. However, that's not to say that rarities do not turn up, and as if to reinforce the point, in late August 2012 Sweden's first Yellow-breasted Bunting for 11 years was trapped and ringed.

Mid- to late October is the best time to see large numbers of Common Eider (*Somateria mollissima*) as they stream past Nabben. Day-counts in excess of 20,000 are not unknown and in autumn 1995 almost 170,000 were counted.

Spotted Nutcracker (*Nucifraga caryocatactes*) of the eastern subspecies *macrorhynchos*. A few nutrackers are noted at Falsterbo most autumns but periodically a large irruption from the east brings hundreds or even thousands.

Seawatching possibilities

Through late autumn it is not just land birds that are on the move. Wildfowl, waders (35 species have been seen), skuas, gulls and terns move past the 'tip' as they empty out of the Baltic Sea. Some seabirds find their way this far east from the North Sea; Leach's Storm-petrel, Sooty and Cory's Shearwaters, for example. The most numerous of the wildfowl are Common Eider (averaging 95,000 a year), Barnacle Goose (averaging 18,000) and Brent Goose (almost 10,000). Other common species, in order of their numbers, are Eurasian Wigeon, Common Scoter, Greylag Goose, Red-breasted Merganser, Northern Pintail, Eurasian Teal, Greater White-fronted Goose, Tufted Duck and Shelduck. Pomarine, Arctic and Long-tailed Skuas are all regular. In September 2007, 68 of the last passed the tip, but that total was exceptionally high.

Spring

The migration of wildfowl and waders into the Baltic Sea is pronounced between March and May, but the visible migration of most near-passerines and passerines is nowhere near as obvious as it is in the autumn, though hirundines are notable exceptions. A better indication of what passerines are passing through in spring is provided by the efforts of the ringing programme, trapping nocturnal as well as diurnal migrants. In late March and early April their nets will contain Meadow Pipit, Winter Wren, Dunnock, Robin, Blackbird, Song Thrush, Redwing, Chiffchaff and other warblers, crests, tits, finches and buntings. Later in April Willow Warblers are plentiful and Tree Pipit, White and Yellow Wagtails, Northern Wheatear, Whinchat, Common Redstart and Thrush

Spotted Nutcracker irruptions

Spotted Nutcracker has a vast breeding range, extending across the taiga forests of high-latitude Eurasia from southern Sweden to Japan, with disjunct populations further south. Its diet comprises mostly the seeds of various pine species. Where these are in short supply, spruce seeds and hazel nuts take their place. Nutcrackers are early breeders, those of the subspecies native to Sweden, *caryocatactes*, for example, laying their eggs in March. Spotted Nutcracker is generally sedentary, not a true migrant, but there is some post-breeding dispersal. Much more dramatic are the periodic irruptions, especially of the eastern thin-billed *macrorhynchos* subspecies, which breeds in Siberia. These occur when there are not enough pine seeds to feed the population. A combination of a poor crop after several successful breeding seasons will generate an irruption. October 1995 witnessed a large movement through Falsterbo, with 1,763 of these beautiful corvids noted, mostly in the period 9–21 October. 1985 was another irruption year and the biggest in living memory was in 1968.

Nightingale will feature along with more of the species already listed. In early May it is the turn of Bluethroat and flycatchers, with Red-breasted being one of the last migrants to feature and Collared being one of the rarer possibilities, along with Eurasian Wryneck and Hoopoe.

The Falsterbo peninsula

A good autumn strategy is to head for Fyren, south-west of the genteel resort of Falsterbo. This is where the old lighthouse is, and it's a good idea first to check out the copse in front of it (return later to see what the ringers are up to). The copse offers cover for birds, and this is a good place to see nocturnal migrants such as warblers and flycatchers the morning after their arrival. Visible migration can be enjoyed from here, but most birders head a few hundred metres further to the very end of the peninsula, Nabben or 'the tip'.

As with all migration sites, the weather determines what birds are moving where – and migration-watchers need to be aware of the options. This is especially true with the birds of prey. Raptor-watching can be at its best when there is a breeze from the south-west or west. Then, the birds will be flying more or less straight into it, so they will seek the shortest, least energy-sapping route to cross the Skjutomrade. That means passing along the southern side of the Falsterbo peninsula, right over Nabben. An excellent place from where to watch is a mound called Kolabacken, close to the southern shore of the peninsula about 1 kilometre from Nabben. If the wind is strong, the raptors will not be flying. If the breeze is from the north-west, most birds of prey will launch south from the peninsula before they reach the tip. Then, the Falsterbo Canal, a few kilometres to the east, will be a better bet. The peninsula is very narrow there so the birds are concentrated into a narrow stream, sometimes quite low.

Yellow Wagtails of the continental subspecies *Motacilla flava flava* ('Blue-headed Wagtail') return to Falsterbo in April. Some breed locally and others pass through. The 'Grey-headed' subspecies *thunbergi* is also regular, as well as many intergrades between the two forms.

Huge movements of Wood Pigeons (*Columba palumbus*) have become more frequent in recent autumns. More than 400,000 were logged by Falsterbo's hard-working counters in each of the 2005, 2006 and 2008 seasons. The biggest day count of all time was 137,500 (give or take a few) on 15 October 2005.

However, there is not as much time to get 'bins on birds' there as a band of trees immediately to the east of the canal limits the skyline. If the breeze is from the south-east, raptors will move on the north side of the peninsula, so Storevang should offer better viewing, and with a north-east tailwind they will head through on a broad front. On warm, sunny afternoons, larger birds of prey often soar over Ljungen heath, midway between Nabben and the canal.

Just north of the lighthouse is the reedbed of South Flommen, where ringing is conducted every autumn. In spring Spotted Crake and Bearded Tit return to breed here, while in autumn the adjacent grassy areas are good for Tawny and Red-throated Pipits and, sometimes, Citrine Wagtail. There are plenty of other sites close at hand, including Slusan and Revlarna (4 kilometres and 6 kilometres north of Nabben, respectively) for waders, and Foteviken (east of the canal) for wildfowl and waders in autumn.

Further afield

If conditions are not good for raptor migration, the 'raptor valley' of Fyledalen, just north of Ystad (to the east), makes a good plan B. Many birds of prey gather here before making the 'final push' if weather conditions aren't optimal. The valley is also good for Black Woodpecker and Hawfinch. East of Lund is Krankesjön, a shallow lake with surrounding reedbeds. This is a 'must-visit' site in spring, when breeding Bittern, Red-necked Grebe, Osprey, Black Tern, Great Reed and Savi's Warblers and Penduline Tits return.

Baltic
Sea
Tallinn
ESTONIA
● MATSALU
BAY
Saaremaa
Pärnu
● SORVE
Gulf of Riga

- **LOCATION**
  Eastern Baltic coast

- **FLYWAY**
  East Atlantic

- **SPRING**
  The first migrants in March are followed by very large numbers of wildfowl and waders, and smaller totals of raptors, skuas, gulls and terns; some passerines are still on the move in late May.

- **AUTUMN**
  Wildfowl; waders; raptors may include Lesser Spotted Eagle; woodpeckers, pigeons, millions of passerines.

- **KEY SITES**
  Matsalu Bay, including Haeska and the Puise peninsula; Poosaspea; Kabli; Sorve on Saaremaa island; in eastern Estonia, Mehikoorma has spectacular visible migration of passerines in autumn.

- **THREATS**
  Matsalu Bay is protected as a national park, but other sites are vulnerable to development.

Matsalu Bay, Estonia

Up to a million waterbirds have been counted passing the western Estonian coast in a single day, and more than 600,000 passerines have been logged in a similar period. The geography of the country makes it the land of visible migration on a truly grand scale.

Estonia is another of those avian crossroads. The northernmost of the Baltic States, it is partly separated from the huge expanse of Russia to the east by the freshwater lake of Peipsi Jarv. To the north, the Gulf of Finland separates it from Finland and the Gulf of Riga is a large inlet of the Baltic Sea to the south. Wildfowl and waders pass along the west or north coasts on their way to and from the northern Baltic or the taiga and tundra of Finland and Russia. Raptors, cranes, owls, woodpeckers, pigeons and passerines breeding in those two countries pass through Estonia in transit. And for many birds it is their final destination in spring.

Given that there is plentiful habitat suitable for use as staging posts or for breeding – shorelines, wet and dry grasslands, reedbeds, fresh and brackish water and forest – it is not surprising that avian diversity is unequalled in northern Europe in spring and autumn. Matsalu Bay National Park covers 486 square kilometres and is one of the most important bird wetlands in Europe. It boasts all the habitats mentioned, including one of Europe's largest wet meadows and 3,000 hectares of reedbeds, and the shallow, brackish, nutrient-rich waters of the bay whose name it bears has a 165-kilometre coastline. It is easy to see why more than 170 species breed here – and why the northern European day-list record of 194 was set here in May 2007.

Harsh conditions
In January and February only the hardiest birds thrive in the frigid conditions. Wildfowl, including Steller's Eider, loaf on open stretches of water,

The flat Baltic coast around Matsalu Bay is a bleak frozen wilderness in winter but with the arrival of spring it welcomes a multitude of wildfowl, waders and warblers, either to breed or on passage to the Arctic.

Common Shelduck (*Tadorna tadorna*) return to the Estonian coast in April, while White-tailed Eagles (*Haliaeetus albicilla*) are residents, predating waterbirds, fish and carrion alike.

with White-tailed Eagles ever on the lookout for food. Owls begin to call and woodpeckers to drum on warmer days but most breeding species are absent. As March progresses there are more and more signs of spring. Northern Lapwings, Skylarks, corvids and Common Starlings move back into the country, with day counts of Skylarks sometimes running into thousands. In mid-April much of the bay is still iced over, but patches of clear water will hold hundreds of Long-tailed Duck and Greater Scaup, with good numbers also of scoter, Common Eider and Smew. During the month Common Cranes fly to forest wetlands, Great Bitterns, Red-necked Grebes and Western Marsh Harriers return to coastal wetlands and passerine migration is in full flow, as it will be through much of May also.

Spring's sudden arrival

The transition from winter to summer is rapid, barely bothering with spring. One week the trees are bare, the next they are in full leaf. Pipits, wagtails (including Citrine), Winter Wrens, Dunnocks, chats, thrushes, warblers, flycatchers and finches all feature prominently. Thrush Nightingale, Blyth's Reed, Barred, Greenish and Icterine Warblers, Red-breasted Flycatcher, Red-backed Shrike and Common Rosefinch are mid- or late-May arrivals.

Raptors and waders are also April and May passage migrants. Some White-tailed Eagles are resident, but others move on to breed in northern Russia. Expected species include harriers, European Sparrowhawk, buzzards, Lesser Spotted Eagle, Osprey and the common falcons. Pallid Harrier, Gyr Falcon and Saker have also been seen in recent springs, with Estonia's first Booted Eagle being found near Poosaspea in mid-May 2010.

71

Thousands of waders stream up the coast, stopping to rest, feed and bathe in wet meadows, at coastal pools or on the shore. Bar-tailed Godwit and Red Knot are in transit to Arctic tundra, Wood Sandpiper and Spotted Redshank to taiga forest, and Ruddy Turnstone and Oystercatcher to northern Baltic shorelines. Sometimes rarities are involved; a few Terek Sandpipers are seen each spring, and two Pectoral Sandpipers were at Sorve in mid-May 2012.

However, it is the country's wildfowl migration for which Estonia is justifiably famed: it is nothing short of sensational, and tens of thousands of geese and ducks are expected daily in late April and early May at several coastal sites. Most numerous are Brent and Barnacle Geese, Long-tailed Duck and Common Scoter; all four species fly east along the Gulf of Finland before passing overland to the White Sea and their Russian breeding grounds. Whooper and Bewick's Swans, Bean and Greater White-fronted Geese, Eurasian Wigeon, Common Eider, Northern Pintail, Common Pochard and Tufted Duck are also plentiful. A few Lesser-white Fronted Geese pass en route to their Norwegian breeding grounds and recent rarities have included Red-breasted Goose, Blue-winged Teal, Ferruginous Duck and Bufflehead.

The nearest breeding populations of Northern Hawk Owl (*Surnia ulula*) are in Finland and Russia. This diurnal predator is not a regular visitor to Estonia but it does turn up in autumn in years when voles have been abundant and the owls have enjoyed a productive breeding season.

Autumn

Wader parties, with Bar-tailed Godwit, Broad-billed and Curlew Sandpipers appear at the coast from late July, and in early August the coast is alive with Northern Lapwing, Wood Sandpiper, Common Greenshank, Common and Spotted Redshank. Ruff, Common Snipe, Eurasian Curlew, Dunlin and Little Stint are plentiful and Temminck's Stint, Red Knot and Red-necked Phalarope can also be found. A vagrant 'yank', Estonia's second Long-billed Dowitcher, was at Parnumaal in September 2012. The first flocks of Common Cranes assemble in August, leaving fields at dusk to spend the night in the relative safety of wetlands. Numbers increase in September before the birds depart for warmer climes.

September and October are months of wildfowl and passerine passage as the northern fringes of Europe drain of their birds before the onset of the big freeze. Among the hundreds of thousands of Goldcrests, Chaffinches, Bramblings and Siskins will be tens of thousands of Jays, thousands of Hawfinches and good numbers of Yellow-browed and Pallas's Warblers from further east. There will also be a smattering of rarities, maybe a Red-flanked Bluetail or a Desert Wheatear. As if to add a final flourish to the birdwatching year, the first flocks of Steller's Eiders, of which 1,500 to 2,500 winter in Estonia's waters, arrive in December.

Working the sites: Poosaspea

North of Matsalu Bay, at the north-west tip of Estonia, is Poosaspea, one of many promontories projecting into the Baltic. Its position makes it ideal for seawatching, and the results are often spectacular, though timing is all-important. In late April and early May birders have described the sea being 'black' with Common Eider, Velvet and Common Scoter and Long-tailed Duck, but they can disappear in a matter of days. Poosaspea is also a top site for *Branta* geese, Black-throated and Red-throated Divers and Red-necked, Black-necked and Slavonian Grebes. Areas of scrub hold migrants in spring and autumn, and a few passage raptors also use the area. Cape Poosaspea is the best place for autumn passage waterbirds. During one season, nearly 2.5 million Arctic-breeding wildfowl passed this narrow bottleneck, with daily totals sometimes exceeding 100,000. On 24–25 September 2011 more than 86,000 Brent and Barnacle Geese were counted, along with almost 13,000 Eurasian Wigeon and 818 Northern Pintail. Up to 2,000 Red-throated and Black-throated Divers have been seen in a day. A Desert Wheatear was another rare find nearby in September 2010.

Puise peninsula

Around Matsalu Bay itself, birders are well catered for with seven tower hides. One of these is at the tip of the Puise peninsula, at the western end of the bay's north shore. The tower gives views south and east into the bay for swans, geese, seaduck and divers. The scrub around the car park can be good for

Sylvia warblers and Red-backed Shrike. By following a track a little north of the car park it is possible to scan the Baltic to the west. Puise is good in spring and autumn. In late September 2012 this site claimed the European record for migrant Lesser Spotted Woodpeckers: more than 180 in a day.

About 5 kilometres east along the north coast of the bay is Haeska bird tower, which overlooks coastal marsh as well as the bay's waters. In spring this is good for geese, dabbling duck, White-tailed Eagle, waders, gulls and terns. It is probably the most reliable place to see Lesser White-fronted Geese, at the end of April and the first week of May, before they move on to breeding grounds in Norway. One group of birders noted 128 species in a day – without moving from this tower! Copses and gardens nearby should be checked for migrants in spring and autumn. In September up to 20,000 Common Cranes gather to feed in fields around the bay before continuing south, and Haeska is one of the best areas to watch them.

Kabli

South of the bay, Pikla Pools and reedbeds are good in spring when they welcome back breeding Great Bittern and other herons, passage Little Gull and marsh terns, breeding Savi's, Great Reed, Reed and Sedge Warblers, and Bearded Tits. South again, Kabli is a classic seawatching spot. Unusually, it is not at the end of a peninsula but the road runs right alongside the coast, so making viewing easy. There is another tower hide here. Kabli is a fascinating place to visit in late September or early October when tens of thousands of birds are caught in Heligoland traps. Here, one can witness bird-ringing and often see birds in the hand, including stunning white-headed *europaeus* race Long-tailed Tits, Willow Tits, Lesser Spotted Woodpeckers and eastern vagrants such as Yellow-browed and Pallas's Warblers. It is worth visiting at first light because Tengmalm's Owls are caught here regularly but they are released early in the morning.

Dunlin (*Calidris alpina*) and Curlew Sandpiper (*Calidris ferruginea*) may stop off at coastal pools before and after making the overland journey to and from the White Sea and their arctic breeding grounds.

Sõrve Bird Observatory

Estonia's largest island, Saaremaa, is reached via a short ferry crossing from Virtsu and a drive across the small island of Muhu. At Saaremaa's southern tip is the famous Sõrve Bird Observatory. A narrow peninsula here points south towards the coast of Latvia. It has a lighthouse, scrubby areas that attract passerine migrants, a lagoon where terns feed and shorelines facing both east and west. Huge numbers of seaduck congregate offshore in late April and early May, and Sõrve witnesses impressive autumn visible migration, especially in the second half of September and October, ranging from thousands of Common Cranes to hundreds of thousands of finches. At least 1,750,000 Chaffinches on 23 September 2010 were followed the next day by 35,000 Siskins and more than 1,000 European Sparrowhawks, the latter no doubt taking full advantage of their travelling companions. In spring 2012 a rare Isabelline Shrike was trapped here, followed by Estonia's fifth Red-flanked Bluetail that autumn. Hundreds of Steller's Eider gather off the north-west tip of the island between January and late March.

Mehikoorma

To the east of the country, spectacular visible migration can be witnessed in late September on the Estonian side of the huge freshwater lake of Peipsi Jarv. At Mehikoorma, just north of Rapina, the lake is at its narrowest where a peninsula extends from the Russian side. This narrowing is particularly favoured by diurnal passerine and near-passerine migrants. For example, between 15 September and 1 October 2012 more than 3 million passerines (mostly finches) were counted, including over 600,000 in a single day. Thousands of Wood Pigeons and good numbers of some raptor species (including Hen Harrier and Eurasian Hobby) use this 'bridge' route in autumn and many water birds, including divers, track the shoreline.

An adult male Pine Grosbeak (*Pinicola enucleator*). This species breeds in the coniferous forests of northern Sweden, Finland and Russia and usually does not move far. However, in autumns when its population outstrips the available fruit and seed supplies, it irrupts. Large numbers move south and west of the beeding range, typically from October onwards. The early winter of 2012/13 witnessed one such movement, with significant numbers observed in transit through Estonia.

- **LOCATION**
  In the German Bight, off the west coast of Germany

- **FLYWAY**
  East Atlantic

- **SPRING**
  Geese and other wildfowl; small numbers of raptors include European Honey-buzzard and Merlin; seabirds, including some local-breeding species; pipits, chats, thrushes, warblers, flycatchers and finches.

- **AUTUMN**
  Seabird passage includes small numbers of shearwaters and skuas; waders; wide range of passerines includes regular Central European and Siberian-breeding species (including genuine rarities).

- **KEY SITES**
  Lighthouse area and Mittelland on Hauptinsel.

- **THREATS**
  Proposed wind farm development.

Heligoland, Germany

Not much more than a pin-prick in the North Sea, Heligoland – the German Fair Isle – is where the scientific study of bird migration began more than 150 years ago. Clearly the island has lost none of its attraction since the days of the great German pioneer Heinrich Gätke.

That migration has been studied on Heligoland more or less unbroken for longer than anywhere else is due in large part to Gätke, who studied birds, and particularly migration, on the island where he lived from 1837 until his death in 1897. He started collecting birds in 1843 and in 1891 he published *Heligoland as an ornithological observatory: 50 years of experience.* This seminal work is important not just for the volume of observations on which it is based but also for the beautiful way it is written. This, for example, is how he describes one chance encounter: "I happened one fine afternoon in May to be standing in front of my house leaning on a low fence, having a talk with an old sailor, when a small bird came flying up the street and settled on the ground between us, so that it almost touched the toes of my boots." The bird was a Baillon's Crake! Most of Gätke's observations concerned the comings and goings of common migrants but he also found some astonishing rarities, including the Western Palearctic's first Black-throated Green Warbler (a first-winter male, in October 1858), from North America, and Eastern Crowned Warbler (in October 1843), from Siberia, the

Only small numbers of Little Auks (*Alle alle*) are present in the German Bight in winter. Sometimes, however, if birds are present in the North Sea, north-westerly winds push hundreds into the waters around Heligoland. This happened on 23 October 2005.

latter event being repeated in October 2012. Building on Gätke's work, Hugo Weigold established one of the world's first bird observatories on the island in 1910, constructing funnel traps, now generally known as Heligoland traps, to assist his bird-ringing operation.

October is a good time to see small numbers of Short-eared Owls (*Asio flammeus*) on the islands.

Pulling power

Heligoland offers *terra firma* for exhausted birds that don't have enough energy to make the European mainland in spring, or meet bad weather on their way west across the North Sea in autumn. It has the same 'pulling power' as Fair Isle or Scilly. Heligoland is actually two islands, 45 kilometres from the German coast in the German Bight. The larger is Hauptinsel, a narrow, rocky triangle with seabird cliffs up to 50 metres high; the other, Düne, is a flatter, sandier, uninhabited affair. With an area of just 1 square kilometre, Hauptinsel is much smaller than Fair Isle, and the largest of the

Scilly Isles, St Mary's, is massive in comparison. Given its history, it's a wonder that Hauptinsel has even survived. It suffered extensive Allied bombing in World War II then – under British occupation – was dramatically reshaped in 1947 by the biggest non-nuclear explosion ever detonated.

The birds

Heligoland has few breeding birds but its checklist runs to 429. Since the main island averages just 500 metres wide and 2 kilometres in length it probably has the largest ratio of migrants to surface area of any patch of land on the planet. Inevitably, these include rarities from all points of the compass. 'Sibes' have included Siberian, White's, Red-throated, Naumann's and Dusky Thrushes and Eastern Crowned Warbler. Black and White-winged Larks, Pine Bunting (1994 and 1995) and Grey-necked Bunting (October 2009, the first German record) are some of the stand-out Central Asian species. Some of the most intriguing records concern southern 'overshoots' and wanderers from continental Europe: Rufous Bush Robin, Cyprus Pied Wheatear (May 1867) and Wallcreeper. Because of the 'blocking' effect of the British Isles, American vagrants are not as prominent as, say, Scilly and this is particularly noticeable with the near-absence of American wood-warblers. However, apart from the previously mentioned Black-throated Green Warbler, a Yellow-throated Vireo found in September 1998 was the second record for

The attraction for tired or weather-beaten migrants of the two islands of Heligoland (Hauptinsel in the foreground, Düne in the distance) is clear.

the Western Palearctic, and other 'Yank' vagrants have included Grey-cheeked and Swainson's Thrushes, American Robin and Grey Catbird. Add a touch of northern flavouring, in the form of Barrow's Goldeneye, Ross's Gull and Hawk and Tengmalm's Owls, and the rarity record looks truly impressive.

Ideal conditions

For seawatching, winds with a westerly component are best, since these will push seabirds further into the German Bight. For most passerines, in spring or autumn, the opposite is generally true so winds with an easterly component may 'drift' many migrants across the eastern part of the North Sea. In 1895 Gätke described this very nicely for nocturnal migration in autumn: "… a calm dark night, without moon or stars, and attended by a very light south-east wind, are the conditions necessary for the grandest possible migration …". If weather conditions deteriorate in the Heligoland area after the birds have set off, then the chances of a major fall on the island become so much better. While this is clearly not an ideal scenario for the migrants, it's a whole lot better than drowning in the North Sea.

One possible spring situation could develop like this. A massive area of high pressure is centred over southern Scandinavia and the Baltic so breezes are moving around the south of it in an easterly direction. As a depression tracks east across southern Britain, the pressure gradient between it and the area of high pressure becomes steeper. In other words, the easterlies become stronger.

The sandy beaches of Düne sometimes play host to Shore Larks (*Eremophila alpestris*) in October.

Flocks of northbound migrants crossing north Germany find themselves being 'drifted' to the west, out over the German Bight. They may be able to readjust and head east, but if the wind is too strong, those whose stamina is failing will find it easier to use the islands of Heligoland as a pit-stop.

Another scenario, this time in autumn, sees flocks of thrushes and finches set off from the coast of Schleswig–Holstein (northern Germany) in light south-easterlies. They head out over the North Sea but after a few hours run into an eastward-moving weather front, with freshening winds and heavy rain. Again, the easiest option for many will be to make landfall on the islands.

Spring and autumn

Spring migration is usually quite an extended affair. Just as Blackbirds often dominate proceedings late in the autumn, so they, along with Common Starlings and Skylarks, are in the vanguard of spring, which on Heligoland is apparent from early March. The first Northern Wheatears appear in the second half of the month, and towards the end of March the first of spring's waves of migration features low-flying lines of Common Scoter heading back towards their Scandinavian breeding grounds, Black-headed and Common Gulls passing offshore and large numbers of White Wagtails. At this time, Woodcock can be flushed from the unlikeliest of places and Chiffchaffs announce their presence with song before moving on to the German mainland. Fulmars, Northern Gannets, Guillemots and Kittiwakes return to their cliff colonies.

During April, thousands of geese pass offshore and on some days there are many Meadow Pipits, Northern Wheatears, Robins, Song Thrushes or Blackbirds. Several Merlins may appear with them. Less numerous Ring Ouzels and Firecrests are regular, Shore Larks are less predictable and the odd rarity – a Bluethroat, for example – shows up. A Short-toed Treecreeper in early April 2012 was the island's 17th record. In the second half of the month the first Eurasian Wrynecks, Reed Warblers, Common Whitethroats and Spotted and Pied Flycatchers show, and Willow Warblers are plentiful. Barnacle Geese continue to pass offshore.

May is an exciting month, with many Common Redstarts and Whinchats added to the parade. The first Icterine Warblers are seen early in the month and sometimes a good passage of European Honey-buzzards is noted around the middle. Even at the end of May there may still be many Common Swifts, Whinchats, Icterine Warblers and Spotted Flycatchers, with lesser numbers of European Golden Orioles and Common Rosefinches and maybe a Subalpine or Greenish Warbler or some other rarity. Marsh Warblers are still moving through Heligoland in the second week of June.

Waders, swifts, pipits and chats

Autumn migration begins in July when the first passage waders are likely to include Sanderling on the sandy beaches of Düne. From the third week,

southbound Common Swifts are noted. The first returning passerines, in mid-August, will probably include Tree Pipit, Northern Wheatear, Willow Warbler, and Spotted and Pied Flycatchers. Seawatching will produce waders, the first small numbers of skuas, and several species of gulls and terns. Although other passerines appear as the month progresses, large numbers are not seen much before the end of the month. Then, Whinchat, Common Redstart and Robin are all likely to be prominent species. Eurasian Wryneck and Barred Warbler are regular but, surprisingly, Red-backed Shrike is not even annual.

With a good westerly blow, seawatching can be productive, with a chance of Sooty Shearwater and Leach's Storm-petrel in addition to four species of skuas. Red-throated and Richard's Pipits are a distinct possibility on Hauptinsel. By late September a whole new suite of migrants begins to dominate, the core species including Meadow Pipit, Blackbird, Song Thrush, Redwing, Fieldfare (with some Ring Ouzels and maybe a Black-throated Thrush), Common Starling, Chaffinch and Brambling. A Little Bunting or Greenish Warbler may turn up along with multiple Yellow-browed Warblers (double figures are possible) and Eurasian Wrynecks if the weather conditions are favourable. Being close to the lighthouse at night can be a memorable experience as the calls of birds drawn to the light are clearly audible.

Brambling (*Fringilla montifringilla*) can be seen in spring and autumn but is more common in the latter season when large numbers may pass with Chaffinches on some days in October. This bird is a female.

All these species increase in October and now scarcer eastern species arrive as well: Richard's Pipit, Yellow-browed, Pallas's and Radde's Warblers and genuine rarities. Blackbirds continue to pour through well into November. And the action does not necessarily end there because a major freeze on the continent will witness new movements, with more geese, ducks, Woodcock and thrushes fleeing across the North Sea.

Hauptinsel and Düne

Heligoland is so small that it is basically two sites, the main island, Hauptinsel, and the smaller Düne. Pretty much the whole of the main island can produce good birds. The bushes around the lighthouse (on the west side) are very good for passerines, especially early in the morning. During nocturnal migration it is worth listening to the calls of birds around the lighthouse. Just north of the lighthouse is the excellent observatory area, but visits need to be arranged. The extensive cover in Mittelland, just south of the lighthouse, can hold very large numbers of migrants. This site has the double advantage of being able to look down on the birds from a relatively sheltered position. This is also a good place for watching visible migration. A few hundred metres the other side of the lighthouse, in the north-east corner of the island, the bushes just south of the Haus der Jugend (youth hostel) can be good for warblers; the sheltered north side of the youth hostel is ideal for seawatching in the right conditions; the bushes around the football pitch nearby attract thrushes, including Ring Ouzels, and the lawns and bushes at Kurgelande can be excellent for pipits, chats, thrushes and phylloscs.

Boats run between Hauptinsel and Düne. Access on the latter is confined to beaches and tracks, and the island does not attract the same range of passerines. The airfield grassland, the area around Golfteich pool and the beach at Dune-Sudostmole are all worth checking for Golden Plover and other waders, larks, pipits, wagtails, Northern Wheatears, other chats and thrushes.

History repeats itself

Eastern Crowned Warbler (*Phylloscopus coronatus*) breeds in the far north-east of China, south-east Siberia, Japan and Korea, with a disjunct population in China's Sichuan province. This sprite winters in southern Thailand, Malaysia, Sumatra and Java. Only a handful of these tiny birds have made it to the Western Palearctic so it seems extraordinary that two have turned up on Heligoland – separated by a gap of 169 years! Heinrich Gätke trapped one on 4 October 1843 and another was seen in the observatory garden on 16 October 2012, further evidence – if any was needed – of the amazing pulling power of the islands to attract vagrants.

FILEY ○ **FLAMBOROUGH**
                ○
Bridlington ●

Yorkshire   North Sea

River Humber
                        SPURN
ENGLAND   ○ **POINT**

- **LOCATION**
  Headlands on the north-east coast of England

- **FLYWAY**
  East Atlantic

- **SPRING**
  Wildfowl, waders, terns, chats and thrushes.

- **AUTUMN**
  Wildfowl; waders; skuas, Little Gull and terns; Short-eared and Long-eared Owls, Eurasian Wryneck; swallows and martins; chats and thrushes; warblers including Barred, Yellow-browed, Pallas's, Dusky and Radde's; Red-backed and Great Grey Shrikes; finches.

- **KEY SITES**
  Spurn Point, Kilnsea 'triangle', Beacon Ponds, Flamborough Head and peninsula; Filey Brigg and ravines.

- **THREATS**
  Most sites are protected but Spurn Point faces constant threat of marine erosion.

Yorkshire coast, UK

Collectively, the Yorkshire coastal sites of Spurn Point, Flamborough Head and Filey have mainland Britain's best rarity record. And this stretch of coast is also great for watching visible migration.

It's a murky early afternoon at Kilnsea in October after a couple of days of light easterlies. Visibility is very poor as you scan the Humber mud. Behind you, quiet trills come from the sycamores outside the Crown and Anchor, and ahead you can hear the distinctive calls of Redwings, though there is no sign of the birds. You wait, and suddenly first one flock, then another, emerges from the gloom: Redwings and Song Thrushes with a few Fieldfares thrown in for good measure. Over the next hour or so a couple of Woodcock and a Hawfinch fly incongruously over a grassy field, Black Redstarts appear at regular intervals and a Great Grey Shrike flies overhead. Then there are the 'crests', hundreds of them – in bushes and hedgerows, on roadside verges and even on the footpath. Back at the Crown and Anchor a 'so-eest' gives away the presence of Yellow-browed Warbler. Perhaps not a typical autumn day on the Yorkshire coast, but this is what it can be like.

Why so good?

Three sites on this coast stand out, quite literally: Spurn Point, Flamborough and Filey Brigg. However, any 'gut', or steep-sided coastal valley, and any patch of cliff-top cover can be a magnet for a tired migrant freshly arrived from continental Europe, maybe a Long-eared Owl, or a Goldcrest or maybe even something rare like a Radde's Warbler. Good birds turn up in such places every year.

The coast offers very good seawatching in spring and autumn, with birds ranging from divers and wildfowl to

A male Common Redstart (*Phoenicurus phoenicurus*). This species, along with Whinchat, Northern Wheatear and assorted Scandinavia-bound thrushes, is regular along the coast in April. Spurn is a reliable site.

skuas and auks. Large numbers of waders move up and down the coast. It benefits from 'drift' migrants in spring and autumn. And it receives hundreds of thousands of passerines from the Continent during the latter period.

Spring

Northern Lapwings, thrushes, corvids and finches are early movers through the Spurn area. In addition to northbound wildfowl and waders following the coast, followed later by skuas and terns, expect to see coasting hirundines, thrushes, corvids and finches in spring. If there is an easterly component to the wind in April or May things could get really interesting, with the possibility of Black Redstart (early), then Eurasian Wryneck, Bluethroat, Icterine Warbler, Red-backed Shrike and Common Rosefinch – in other words, birds that have drifted west of their intended course to continental Europe.

Autumn

Although good birds are found every spring, the Yorkshire coast really shines in the autumn. Return wader passage starts in July, with sites such as Tophill Low, Hornsea Mere, Beacon Ponds and Easington Lagoons (between Kilnsea and Easington) benefitting. In August and September it gathers steam and Yellow Wagtail, Common Redstart, Whinchat and Northern Wheatear feature prominently, especially in the Spurn area.

Radde's Warbler (*Phylloscopus schwarzi*) breeds in southern Siberia and winters in South-east Asia. However, presumed 'reverse migrants' are regular on England's east coast in autumn, particularly in late September and early October, sometimes in reasonable numbers.

The end of August and early September is also prime time, particularly if there has been an easterly component to the wind, for Eurasian Wryneck, Barred Warbler and Red-backed Shrike. For example, if, after a period of westerlies, the wind switches to the east in late September or October, expect scarce migrants to appear as if from nowhere. Corncrake, Barred and Yellow-browed Warblers, Red-breasted Flycatcher, Red-backed Shrike and Common Rosefinch are all good shouts. In these conditions there may be scores of Northern Wheatears and Common Redstarts at the best sites. Double-figure counts of Yellow-browed Warbler are no longer unthinkable.

Offshore, Northern Gannets, skuas and terns trickle – then stream – along the coast. If there is a northerly or easterly component in the wind, Filey Brigg, Flamborough Head or Spurn Point could witness excellent seawatching, including Sooty Shearwaters, Pomarine or even Long-tailed Skuas. Even without this it is worth keeping an eye on the North Sea. Little Gulls can sometimes be present in large numbers. For example, in early October 2012 an estimated 5,000 were feeding off Spurn, slowly moving into a light north-westerly. Wildfowl include Whooper Swan, Pink-footed and Brent Geese, Eurasian Wigeon and Common Scoter.

Vis-migging spectacular

Thousands of Song Thrushes, Blackbirds, Redwings and Fieldfares – and smaller numbers of Ring Ouzels – pour in, along with many Meadow Pipits, Goldcrests, Chaffinches and Bramblings. This makes for sometimes spectacular 'vis-migging'. Inevitably, Richard's Pipits are found every September and October. Short- and Long-eared Owls can be seen coming in off the sea – a marvellous sight. A classic scenario unfolded on the morning of 22 October 2012 when, after a wet night, the day dawned foggy with a light north-easterly. Thrushes and Bramblings had been coming in off the North Sea overnight and this influx continued apace until late morning, with lighter passage in the afternoon. At Spurn, counts of 21,100 Redwing and 2,675 Brambling were new records but they did not tell the whole story as there were also dozens of Ring Ouzels, hundreds of Robins and Goldcrests and thousands of Blackbirds, Fieldfares and Song Thrushes in the mix.

At this time, scarce and rare warblers appear in coastal cover. Nowadays, multiple arrivals of Yellow-browed Warblers can be expected at the main coastal sites (especially Filey's ravines, Scarborough Castle, Flamborough's South Landing and Kilnsea), with smaller numbers of Pallas's following a little later. Radde's and Dusky Warblers are also annual. October is also the time for Red-breasted Flycatchers and – in recent years – Red-flanked Bluetails, which are now approaching annual on this coast. Little Gulls stream south, offshore, and if there has been a dislocation of Little Auks into the North Sea, November is the month when good numbers are likely to be seen from Filey Brigg and Flamborough Head; as with other seabirds, their movements are dependent on winds.

Long-eared Owl (*Asio otus*) is sometimes seen coming in off the sea in October. This species and Short-eared Owl (*Asio flammeus*) can then turn up anywhere, and tired migrants are often disturbed from the ground unintentionally.

Males of the northern race of Bullfinch (*Pyrrhula pyrrhula pyrrhula*) are bigger and brighter than their British counterparts. In some autumns birds move south-west from northern Europe and a few reach Britain's east coast. In October 2004, for example, there were about 10 at Flamborough.

Spurn Point and the Kilnsea 'triangle'

Spurn Point is a long and very narrow spit, the product of longshore drift across the northern side of the Humber estuary. It juts further into the North Sea than anywhere between Gibraltar Point to the south and Flamborough Head to the north. Birds that are working down the coast from the north become channelled into a narrowing 'funnel' (reminiscent of Cape May in the United States), and unobstructed skylines offer great views of visible migration. It often pays to walk the Kilnsea 'triangle' from the Crown and Anchor to the churchyard just to the east and then along the road to Beacon Lane, checking the fields either side, and the ditch to the south. By the Bluebell Inn, turn south to the bird observatory and a hide overlooking a pool. Return along the Humber seawall or come back the way you have walked, continuing up Beacon Lane, which has thick hawthorn and buckthorn cover either side for much of the way to a hide overlooking Beacon Ponds. The lighthouse, elders and sycamores of Point Camp, at the tip of the hook-tipped Spurn peninsula, are 4 kilometres beyond the observatory.

The site's catalogue of rarities includes: Oriental Turtle Dove (in November 1975); a long-staying Tengmalm's Owl (March 1983); Chimney Swift (August 2000); Pacific Swift (July 2005 and June 2008); Calandra Lark (October 2004); Britain's first Black Lark (April 1984), though it was not identified as such at the time; Cliff Swallow (October 1995); Pallas's Grasshopper Warbler (late September 2008); Pechora Pipit (September 1996); Blyth's Reed Warbler (May 1984), the first British bird of this species to draw a crowd; Asian Desert Warbler (May 2000); and a singing male Marmora's Warbler (8–9 June 1992).

The Flamborough peninsula

There is much more to Flamborough than the Head. It sits at the end of a large, roughly triangular peninsula, much of it cultivated and inaccessible, jabbing out 10 kilometres into the North Sea. There are cliffs of varying heights to north and south, broken every so often by uncultivated 'guts', the first port of call for any migrant hunter. The interior of the peninsula is a patchwork of fields, broken up by copses and hedgerows. Fields attract large numbers of thrushes and Common Starlings. Where there are stubbles big flocks of finches and buntings can congregate. The wetter areas are favoured by wildfowl. In an area with such limited cover, it is little wonder that tired migrants seek refuge in the patches that are available. A footpath runs around the peninsula, so fans of seawatching can vary their position according to the prevailing wind. The simplest option is to park near the Fog Signal Station at the tip. There are masses of brambles in this area, which can be alive with migrants.

South of Flamborough village, the wooded valley leading to South Landing always produces good birds in autumn, Red-flanked Bluetail in October 2011, for example. Between here and Flamborough Head, running south–north from the coast path to the B1259 is Old Fall Hedge, with tiny Old Fall Plantation set off to the east. Britain's first Asian Brown Flycatcher was found here on 3 October 2007. North-east of the village, near the Viking pub, is an excellent bit of habitat: the scrubby Holmes' Gut, which is a draw for Yellow-browed and Pallas's Warblers and 'eared' owls. Park carefully on the road and walk north past the pumping station. Flamborough 'megas' have included Pallid Swift (July 1992); Rufous Bush Robin (October 1972); Asian Desert Warbler (October 1991); Britain's first Taiga Flycatcher (trapped in April 2003); and Brown Shrike (September 2008).

Yorkshire did not receive its first Red-flanked Bluetail (*Tarsiger cyanurus*) until 2007, but there have been several since, including two on one day at Spurn in mid-October 2009. The range of this magical chat has been extending west in recent years and now breeds as close as Finland. It winters in southern China and South-east Asia.

The Filey area

Several sites compete for birders' attention around Filey. For seawatching, the narrow finger of low rocks called Filey Brigg is the place to be. South of the Brigg, the beach can be interesting for waders and wildfowl and two steep-sided narrow valleys, Arndale Ravine and Church Ravine, are guaranteed to have phylloscs when they are on the move. The country park, above the ravines, has open grass for pipits and scrubby areas where birds such as Radde's Warbler have been found. The 'top field' is good. Pechora Pipit (October 1994), a singing Spectacled Warbler (late May 1992) and a Two-barred Greenish Warbler (October 2006) are a few of the site's most impressive 'rares'.

Celtic Sea ENGLAND

Cornwall

PORTHGWARRA ○

ISLES
OF SCILLY ○

English Channel

Scilly and West Cornwall, UK

- **LOCATION**

  Islands and headlands in extreme south-west of England

- **FLYWAY**

  East Atlantic

- **SPRING**

  Terns, hirundines, chats, thrushes, warblers and flycatchers; regular scarce birds include European Bee-eater, Hoopoe and Woodchat Shrike.

- **AUTUMN**

  From late July migrant waders and seabirds, including Balearic, Cory's, Great and Sooty Shearwaters; then chats, thrushes, warblers, flycatchers and buntings. American waders are possible in August and likely in September; other Nearctic vagrants, including Red-eyed Vireo and wood-warblers, most likely in October; eastern scarcities, including phylloscs, from mid-September to November.

- **KEY SITES**

  Lower Moors, Porth Hellick Pool, airfield and The Garrison on St Mary's; St Agnes, Bryher and Tresco Great Pool; Gwennap Head, Pendeen Watch and St Ives headlands; Porthgwarra, Cot, Nanquidno and Nanjizal Valleys.

They may not provide the visible migration spectacle of some other hotspots, but this collection of small islands and gale-battered headlands is the rarity finding zone par excellence.

There is something enormously exciting about approaching the Cornish town of Penzance in late summer or autumn. Every way you turn the opportunities for amazing birdwatching abound. In August, nearby Marazion Marsh may hold a Spotted Crake or Aquatic Warbler and there could be an American wader feeding among the seaweed on the beach. Just up the road, the Hayle estuary and Drift Reservoir will have their first instalments of autumn wader migrants. Depending on the synoptic chart, a trio of headlands – Porthgwarra, Pendeen and St Ives – could deliver memorable seawatching experiences. A little later in the season and a short boat-trip away, the magical Isles of Scilly beckon.

The Isles of Scilly

Lying 45 kilometres off Lands End, this collection of beautiful islands, some small and the others tiny and uninhabited, has for many years been regarded as Britain's 'rarity central' and for good reason. Since the 1950s the islands have produced a remarkable catalogue of rare and scarce birds from west and east, including many UK firsts. More than 420 species have been recorded on islands with a total area less than the London borough of Westminster.

The attraction is clear. For birds that have flown too far north in spring or too far west in autumn, these are the last chunks of rock before the Atlantic. Any tired, lost bird that finds itself over the ocean has only two choices: to drown or head for the profile or lights of the inhabited islands of St Mary's, St Agnes, Tresco, St Martin's or Bryher. In this respect Scilly has the same kind of migrant-pulling power as Heligoland or Fair Isle.

Additionally, Scilly has amassed an extraordinary array of New World vagrants that have been caught up in fast-track weather systems crossing the Atlantic in autumn and made some of the most remarkable journeys imaginable. Some, particularly waders but also passerines weighing just a few grams, will have made it under their own wing-power. Others may have hitched a lift on a boat for at least part of the way but even their journeys are remarkable. In this respect Scilly has much in common with Cape Clear Island in Ireland.

Vagrant nearctic passerines usually disappear a few days – if not hours – after they are discovered. Not so with this Northern Waterthrush (*Parkesia noveboracensis*), which was found at Lower Moors, St Mary's, in September 2011 and remained faithful to the area for the next seven months. This species breeds in forests in Canada and New England and spends the winter in mangroves in the Caribbean and northern South America.

The result is a remarkable mix of common, scarce and rare migrants, ranging from seabirds to waders and passerines. This mix is perhaps best illustrated by an example from birding folklore. On 7 October 1999 two rare species of American waders – Upland Sandpiper and White-rumped Sandpiper – were on the largest of the islands, St Mary's. Those lucky enough to have seen them travelled over to St Agnes, where a 'mega' from the east, a White's Thrush, had been playing hard to get and a first-winter male Siberian Thrush was on neighbouring Gugh. While birdwatchers were waiting for the latter to appear, what turned out to be Britain's first Short-toed Eagle drifted overhead. This should have been migrating from southern Europe south into Africa.

Fewer Nearctic passerines seem to make it to Scilly these days and there has been much speculation that this is due to a combination of population declines of many North American forest species, a more northerly track for autumnal depressions and fewer birders on the islands. Be that as it may, some still make it, as the Buff-bellied Pipit and Blackpoll Warbler in October 2012 showed. And the islands still draw new birds. In the same month, a Sykes's Warbler from Central Asia was on Tresco – a first record for Scilly.

European Bee-eater (*Merops apiaster*) is possible on Scilly in spring and autumn. This bird was at Newford Duck Pond, St Mary's, in October 2011.

The western extremity of Cornwall benefits in much the same way, though the rarities are never as concentrated. A series of short, narrow, steep-sided valleys around the coast of the Penwith peninsula are not suitable for agriculture and their uncultivated slopes have just the kind of cover that tired migrants are desperate for. The valleys are sometimes hard to work and do not receive the amount of attention devoted to the lanes and paths of Scilly – but they have nonetheless produced many rare birds. Additionally, the nearby Hayle estuary and Drift Reservoir are wader-traps, but it is the seawatching on offer from the south-west and west-facing headlands of Porthgwarra, Pendeen and St Ives for which west Cornwall is most famous.

Spring 'overshoots'

A Barn Swallow or Northern Wheatear is sometimes seen at the end of February in this, the south-west extremity of England, but spring migration really begins in mid-March, maybe with the sighting of a few Sandwich Terns off a headland or the appearance of a party of Northern Wheatears on St Mary's airfield or Porthgwarra. A few Whimbrel turn up on the coast and Sand Martins appear as if from nowhere. Later in the month, Chiffchaff and Blackcap song is heard, Barn Swallows appear in good numbers and a few Black Redstarts and Ring Ouzels are found. In early April wader passage becomes more obvious, with Common Greenshank dropping in at Tresco Great Pool, Common Terns offshore, the first House Martins seen and Willow Warblers' tuneful cadence heard in country lanes. There is a good chance that a southern 'overshoot' – maybe a European Bee-eater, Hoopoe, Serin or Woodchat Shrike – will be found. There were five of the last species on St Mary's in mid-May 2011. Where else in England could that happen? Northern Wheatears and Meadow Pipits are often on any patch of short coastal turf, Common Redstarts and Common Whitethroats appear, and a light passage of raptors is noted, including a handful of Western Marsh Harriers, Ospreys and Merlins. A few Short-eared Owls are likely.

Rare herons

Scilly has a reputation for attracting scarce herons in spring, and Spoonbill or Purple Heron are possible; there have been multiple arrivals of Black-crowned Night Herons. As April moves on, Common Cuckoo, Tree Pipit, Whinchat, Lesser Whitethroat and Spotted Flycatcher stop over. Neither Scilly nor Penwith are famous for large numbers of passerine migrants in spring but every year is different and small falls do occur. At the end of April 2012 there were "many hundreds" of Northern Wheatears, Blackcaps and Willow Warblers on Scilly and an extraordinary 24 Grasshopper Warblers on Bryher alone. A few Common Swifts pass through at the end of April and the start of May.

Early May is an exciting time with a continuation of wader passage and an opportunity to find birds that have strayed too far north. These include Hoopoe, Red-rumped Swallow, Greater Short-toed Lark, Woodchat Shrike and European Golden Oriole. Scilly has an especially good record for Red-rumped Swallow and European Golden Oriole and there may be half a dozen of each on the islands during May. On 12 May 2012 there were thought to be six of the former on St Mary's; and six days later there were half a dozen orioles on the same island. Small numbers of Turtle Doves and Spotted Flycatchers may be seen any time during the month and there is a good chance of a Common Quail, Nightjar, Common Nightingale or even a European Bee-eater. The last stragglers of the spring migration are still evident in the first half of June, not long before the return migration of waders begins.

The 'Scilly season'

The Penwith headlands offer some of the best seawatching in Britain from July onwards. Those on Scilly do not have quite the same reputation but short-distance pelagic trips around the islands can deliver. If conditions are good, Wilson's and European Storm-petrels, Cory's, Great, Sooty, Balearic and Manx Shearwaters, and Great and Arctic Skuas are all possible on pelagic trips or

A classic autumn

In a short period of late September and October 1999 Scilly hosted a Short-toed Eagle (from continental Europe), no less than nine White-rumped Sandpipers, two Upland Sandpipers, two Common Nighthawks and a Baltimore Oriole (from North America), and White's and Siberian Thrushes and nine Radde's Warblers (from Siberia), Just across the water on the coast of west Cornwall, there were Yellow-billed Cuckoo, Chimney Swift and Veery (from North America), a Blue Rock Thrush (from southern Europe) and a Citrine Wagtail from eastern Europe. And there were lots of scarce, rather than rare, birds as well.

from land-based seawatches in late July and August. Returning waders begin to appear on beaches and around waters such as the Great Pool on Tresco. Common Swift is in the vanguard of the birds returning to sub-Saharan Africa and Green and Common Sandpipers, Whimbrel, Black-tailed and Bar-tailed Godwits and Common Greenshank arrive.

Mid-August is the best time for Pied Flycatchers, which can appear anywhere on the islands or in the Penwith valleys of Porthgwarra, Kenidjack, Cot and Nanquidno. Towards the end of the month the first American waders can appear; these are most likely to be Pectoral Sandpiper, Buff-breasted Sandpiper and American Golden Plover. In total, 18 species of Nearctic waders have been recorded on Scilly alone. Scarcer species among the Palearctic waders include Little Stint and Curlew Sandpiper.

By the end of the month common migrants are found, including Tree Pipit, Northern Wheatear, Whinchat, Common Whitethroat, Willow Warbler and Spotted Flycatcher. Late August and early September is a good time for scarcities such as Eurasian Wryneck, Melodious and Barred Warblers, and Red-backed and Woodchat Shrikes, especially in the Cornish valleys. Spotted Crake is possible at Porth Hellick Pool, and Garganey at Tresco Great Pool. Although the number of birds is rarely very large, the variety in September is impressive. Scilly scores consistently with several species that are rare elsewhere. While British Ortolan Bunting numbers have decreased in tune with the species' decline as a breeding species on the Continent, Scilly still punches above its weight. In September 2012 there were at least three on Tresco and at least one on St Mary's. Other regular scarcities and rarities include Corncrake, Richard's Pipit, Citrine Wagtail, Yellow-browed Warbler, Rosy Starling, Common Rosefinch and Lapland Bunting.

October is the month for which Scilly is rightly famed, the month when birders might find themselves looking at a Red-eyed Vireo within minutes of watching a Red-throated Pipit, or a Grey-cheeked Thrush and an Olive-backed Pipit on the same morning. The period from late September through

October can be brilliant but, as always, much depends on the weather locally and further afield.

Apart from the 'megas' this is the prime month for American waders, Red-throated Pipit, Booted, Hume's, Yellow-browed and Radde's Warblers, Red-breasted Flycatcher and Little Bunting. The supporting cast is likely to include Black Redstart, Ring Ouzel, Brambling and Lapland Bunting, and Grey Phalarope, skuas and Sabine's Gull offshore. This is also a good time for more surprising Scilly rarities, maybe a Bullfinch or even a less-than-annual Coal Tit! Migrants continue to arrive in November, almost exclusively from the east, now. The first Woodcock, Jack Snipe and – if it is an invasion year – Waxwing are seen. Even towards the end of the month there may be influxes of Blackbirds, Fieldfares and Redwings. And, of course, if there is a big freeze in northern or eastern England, expect a fresh invasion of mainland species desperate for food.

Yanks and Sibes

Britain's first Great Blue Heron was seen at Lower Moors, St Mary's, for just one day in December 2007. The only British Magnolia Warbler was on St Agnes in September 1981, a Philadelphia Vireo on Tresco in October 1987 holds the same distinction and the same month witnessed the discovery of

Snow Bunting (*Plectrophenax nivalis*) is an autumn passage migrant and winter visitor in small numbers. Porthloo Beach, on the west coast of St Mary's, is a place worth checking for this species.

Britain's only Wood Thrush, on St Agnes. There is no room to list all the trans-Atlantic species but some of the other exciting fare has included Wilson's Snipe, Chimney Swift, Tree and Cliff Swallows, Hermit Thrush, Scarlet Tanager, several Northern Waterthrushes and an Andrex-long list of wood-warblers.

Being on the 'wrong' side of Britain, Scilly and west Cornwall do not have the same pedigree when it comes to adding eastern birds to the British List, but that in no way detracts from the quality of species seen. Caspian Plover, Bimaculated Lark, Rufous Turtle Dove (there are records from Scilly and Cornwall), Siberian, Eye-browed and White's Thrushes, Olive-backed, Blyth's and Pechora Pipits, and Blyth's Reed, Sykes's, Booted and a whole suite of *Phylloscopus* warblers are just a selection.

St Mary's

Boats and flights come in to St Mary's, the biggest of the Scilly Islands. It is small but has a good mix of habitats. Just west of the 'capital', Hugh Town, is The Garrison, a headland with pine and elm trees and a playing field. All kinds of migrants have been seen here, including many rarities. Britain's only Green Warbler was found here in late September 1983. On the other side of Hugh Town are the pools and reedbed of Lower Moors, where Britain's only Great Blue Heron, another American vagrant, was in December 2007. It is good for crakes, rails (there was a Sora here in 2005) and Jack Snipe, and it hosted a long-staying Northern Waterthrush in 2011.

To the south is Peninnis Head, probably the best seawatching location in the archipelago. The road to it is planted with seed crops to attract finches and buntings, the grassy areas near the headland are excellent for pipits and Northern Wheatear and among a fine collection of shearwaters, skuas, gulls and terns seen from Peninnis itself have been some pelagic gems, including a Black-browed Albatross on the sea in September 2009. The airfield attracts waders such as Dotterel, Buff-breasted and Upland Sandpipers and American

The Turk's Head on St Agnes is a welcome port of call for birders after an exhausting – or exhilarating – session in the field.

Golden Plovers, pipits, Ortolan and Lapland Buntings, Eurasian Wryneck and Greater Short-toed Lark. Just beyond the airfield is a cluster of productive sites, Porth Hellick Pool and reedbed, Higher Moors and the dense cover of Holy Vale. Even away from the best sites, many of the island's lanes are lined with trees (especially elm), which attract phylloscs, Firecrests and Red-breasted Flycatchers in autumn. Between the lanes are cattle-grazed fields (good for larks and pipits) or bulb fields, separated by hedges but with a network of paths to explore.

St Agnes and the other islands

Just 2 kilometres from north to south, St Agnes is the most south-westerly of the inhabited islands. St Agnes has many claims to fame, including British 'firsts' for Northern Waterthrush (1958), Magnolia Warbler (1981), Wood Thrush (1987) and Short-toed Eagle (1999). Much of the island is a patchwork of fields separated by hedges, which can be good for concealing migrants. The trees at The Parsonage, near the lighthouse, represent the most famous spot on the island. This is good for migrant passerines – including rarities – in spring and autumn. Leading south towards the heathland of Wingletang Down is Barnaby Lane, where Britain's only Magnolia Warbler was found in October 1981. And Britain's first Cream-coloured Courser for 20 years was found at Wingletang Down in 2004; this is also a good location for Richard's Pipit. Paths lead over the down towards Horse Point, at the southern extremity, a good seawatching spot in the right conditions. The beaches at Periglis and Porth Killier are the best places for waders. To the east, St Agnes is connected to even smaller Gugh by a sandbar at low tide. Gugh has its own claims to fame, including Britain's first Short-toed Eagle.

Tresco does not have quite the same rarity pedigree as St Mary's or St Agnes but, even so, not many locations can compare with it. Britain's first Northern Parula in 1966 and Philadelphia Vireo in 1987 bear testament to that. The Great Pool is the largest area of fresh water on the islands and has hosted Nearctic wildfowl such as American Black Duck, Ring-necked Duck and Blue-winged Teal. It also has a reedbed and fields to the north, which can hold finches and buntings. To the west of Tresco, Bryher has scored more than its fair share of rarities over the years (including Common Yellowthroat in 1984, and Common Nighthawk and Baltimore Oriole in 1999), but neither this nor St Martin's, the last of the inhabited islands, attract as many birdwatchers as they deserve.

The Cornish headlands

A narrow lane with hedgerows either side leads to the bottom of the valley at Porthgwarra. On occasions in autumn is has not been necessary to leave the car park to see good birds, Yellow-browed and Dusky Warblers and Red-eyed Vireos included. Even if there are no rarities, this is a good place for scarce and common passerines in spring and autumn. Open areas attract

pipits and Northern Wheatears, while bushes and scrub hold warblers, flycatchers and the odd shrike. Above the valley and looking out into the South-West Approaches, Gwennap Head is one of the UK's top seawatching sites, particularly if a south-westerly is blowing. It is particularly good for shearwaters between late July and early September. During this period, Manx can be expected and there's a good chance of Balearic and Sooty, with a half-chance of Cory's and Great Shearwaters; sometimes the last two species go through in big numbers but these movements are hard to predict. In late July 2009 a Black-browed Albatross was seen from here, with a Fea's-type Petrel just five weeks later.

The coastal cliffs between here and St Ives are periodically punctuated by narrow, steep-sided valleys, which are unsuitable for grazing and often have thick cover. All are worthy of investigation in spring and autumn because they are migrant-traps, but most are hard to work and sections are inaccessible. Many rare birds have been found among the commoner migrants, but one can only speculate on what has visited unnoticed. Just east of Porthgwarra, St Levan had a Veery in October 1999; just to the west, Nanjizal hosted Britain's first Alder Flycatcher in October 2008; migrants find the tree cover at Sennen Cove, just north of Land's End, impossible to resist. Britain's first Bay-breasted Warbler was found nearby in 1995. Thirteen years previously, one of the most extraordinary rarity discoveries was made in the Nanquidno Valley: a Varied Thrush, which should have been on the west coast of North America; this remains the sole Western Palearctic record. Nearby, St Just airfield is worth scanning for Buff-breasted Sandpiper and Dotterel in autumn, and larks, pipits, wheatears and thrushes in either migration. Cot Valley hosted both Yellow-billed Cuckoo and Blue Rock Thrush in autumn 1999 and, more

Jack Snipe (*Lymnocryptes minimus*) is a regular autumn visitor to the pools at Porth Hellick and Lower Moors, St Mary's. Passage birds are sometimes found on local beaches.

recently, a White's Thrush in October 2012, the first in Cornwall for half
a century; and Britain's only Yellow-throated Vireo was at the next valley,
Kenidjack, in October 1990.

Just north again, Pendeen Watch produces its best seawatching with
westerlies after a period of south-westerlies. Autumn is best for shearwaters,
storm-petrels, Grey Phalarope and skuas. St Ives Head is the third of the trio
of seawatching promontories. Both Porthgwarra and Pendeen have overhauled
its reputation in recent years, but those who were there in early September
1983 will never forget it. After a period of south-westerlies a very strong
westerly gale proceeded to deposit large numbers of seabirds in St Ives Bay.
During the day 20,000 Northern Gannets, 25,000 Manx Shearwaters, 10,000
European Storm-petrels comprised the majority, but 200 Sabine's Gulls, more
than 50 Great and 30 Balearic Shearwaters, four species of skua, 10 Leach's
and one Wilson's Storm-petrel completed the picture.

**For close views of
Sabine's Gulls (*Larus
sabini*) – this is a
juvenile – an autumn
pelagic trip into the
waters off Scilly is
more likely to deliver
than land-based
seawatching.**

County Cork
IRELAND
Baltimore

CAPE CLEAR ISLAND

Celtic Sea

- **LOCATION**
  South-west coast of Ireland

- **FLYWAY**
  East Atlantic

- **SPRING**
  Skuas, terns, hirundines, chats, warblers; regular scarcities include Turtle Dove, Hoopoe and Eurasian Golden Oriole.

- **AUTUMN**
  Seawatching for shearwaters, storm-petrels and skuas can be superb in July and August; waders, including the chance of an American species; pipits, chats, thrushes, warblers, flycatchers and buntings; regular scarcities include Eurasian Wryneck, Barred and Yellow-browed Warblers and Red-backed Shrike; American and Siberian vagrants are possible.

- **KEY SITES**
  Blannan, Cotter's Garden, East and West Bogs; on the near mainland there are other good seawatching headlands and good sites for waders.

- **THREATS**
  None.

Cape Clear, Ireland

If you like the idea of combining top-class seawatching with some find-your-own birding, then the beautiful island of Cape Clear will probably appeal.

Cape Clear Island is to the Irish mainland what Scilly is to Britain – a first port of call or the end of the road, depending on the provenance of the migrant in question. It has claimed Irish 'firsts' for species as varied as Black-throated Diver, Bulwer's and 'soft-plumaged' Petrels, Black-browed Albatross, Little and Chimney Swifts, White-throated Needletail, Grey Catbird, Siberian Thrush, Zitting Cisticola, Blyth's Reed, Sykes's and Pallas's Warblers, no fewer than six different species of American wood-warblers and Indigo Bunting. The make-up of that list, which is not definitive, comprises 'Yanks', 'Sibes' and oceanic wanderers. In addition, Cape Clear regularly attracts species that are scarce in Ireland, from Turtle Dove to Common Nightingale, Reed Warbler to Lesser Whitethroat and Wood Warbler to Yellow-browed Warbler.

Cape Clear Island is a hilly outcrop about 5 kilometres long and two kilometres at its widest. About 8 kilometres from the nearest mainland port at Baltimore, this chunk of old red sandstone is part barren moorland, part bog and part fields, with some mature gardens and small plantations.

The bird observatory was established in 1959, and some golden years of seawatching and rarity-finding followed, the latter exemplified by a series of finds by North Harbour on 7 October 1962: in the same patch of bushes were Europe's first Rose-breasted Grosbeak, a Subalpine Warbler and a Red-backed Shrike!

Why so good?

The hilly island points south-west into the Atlantic, almost an extension of one of County Cork's many long, narrow peninsulas.

Wilson's Storm-petrel (*Oceanites oceanicus*) breeds in the southern hemisphere but outside the breeding season pursues a trans-equatorial migration. It is an outside possibility from Blannan headland in August.

Long-tailed Skua (*Stercorarius longicaudus*) is a rarity in Ireland but a few are seen from the headlands of the south-west in autumn. Just four were seen from Cape Clear Island in August and September 2012.

That means its southern headland of Blannan is in an ideal position for seawatching at pretty much any time of year, but especially in late summer when Great and Cory's Shearwaters track south through the eastern Atlantic and in autumn when post-breeding Northern Gannets, Sooty Shearwaters, storm-petrels, Grey Phalaropes, skuas and Sabine's Gulls follow their flight paths. Blannan is, of course, not the only seawatching headland in south-west Ireland. Others in Cork, Kerry and Clare have come to prominence in recent years, including Bridges of Ross, the Old Head of Kinsale, Mizen Head and Galley Head, but Blannan still produces fine birds and spectacular numbers.

Ideally placed

Being 250 kilometres west of St Mary's, Scilly, and 150 kilometres north, it is ideally placed to catch Nearctic vagrants that have got themselves caught up in autumn's vigorous transatlantic depressions. This explains the presence of Yellow-billed Cuckoo, Yellow-bellied Sapsucker, Chimney Swift, Grey Catbird, Hermit, Swainson's and Grey-cheeked Thrushes, Red-eyed Vireo, Black-and-white, Blue-winged, Yellow, Yellow-rumped and Blackpoll Warblers, American Redstart, Northern Waterthrush, White-throated Sparrow, Rose-breasted Grosbeak, Indigo Bunting, Bobolink and Baltimore Oriole on the island's checklist. And then there are the American waders, although there are much better mainland sites for those. It is strange to think that while

Yellow-bellied Sapsucker has been recorded, neither Green nor Lesser Spotted Woodpecker has. Other notable absentees are Marsh and Willow Tits and Nuthatch. And in much the same way that Scilly mops up eastern vagrants that would otherwise end up drowned at sea, so Cape Clear performs much the same role. It played host to Ireland's first Siberian Thrush and Pallas's Warbler, the country's first Sykes's Warbler five years later and Blyth's Reed Warbler in 2006. Yellow-browed Warbler is annual in autumn; indeed, there were estimated to be at least 50 on the island in the third week of October 1985.

Then there are the southern 'overshoots', which have included Eurasian Scops Owl, Little Swift, European Bee-eater, Red-rumped Swallow, Ireland's first Rufous Bush Robin, Hoopoe and Golden Oriole.

Spring and summer
March arrivals include Sandwich Terns, Sand Martins, White Wagtails, Northern Wheatears and Black Redstarts, Ring Ouzels and good numbers of Chiffchaffs.

Turtle Dove (*Streptopelia turtur*) is more likely in spring than autumn. A handful can be expected in the second half of May and early June.

The main rush takes place in April, with Common and Arctic Terns passing offshore, hirundines (possibly including a Red-rumped Swallow), pipits, Yellow Wagtails and a variety of warblers. The last group may include a few Reed Warblers and Lesser Whitethroats, scarce birds in Ireland, and a rarity or two. Early May is the best time to look for Eurasian Hobby, Turtle Dove, Spotted Flycatcher and Eurasian Golden Oriole. The last species is annual in spring, but don't expect a repeat of May 1994 when there were 14 on the island.

By mid-July it is time to switch attention to seawatching. With onshore winds large numbers of Manx Shearwaters may be passing. Particularly from the last week of the month and into August, in the right conditions it should be possible to pick out Sooty, Cory's and Great Shearwaters (sometimes in large numbers, though numbers vary from year to year, as well as with the wind), Great, Pomarine and Arctic Skuas, European Storm-petrels and terns. Now is the time to check every passing storm-petrel for a Wilson's, since more are being identified from *terra firma* these days, and every gliding shearwater for a Balearic or even a Fea's-type. Seawatching interest will continue right through the months of autumn.

Autumn

Late August witnesses the arrival of small waves of common migrants such as Northern Wheatear, Common Whitethroat, Blackcap and Willow Warbler. Among them may be a few Pied and Spotted Flycatchers and this is a good time of autumn to find a Eurasian Wryneck, Barred Warbler or Red-backed Shrike, which are all fairly regular at this time. The Wheatear Field may attract a Buff-breasted Sandpiper in late August or a Lapland Bunting in September. September and October should witness the visible migration of pipits, wagtails and finches, sometimes including Common Crossbills. Chats, including a smattering of Black Redstarts and Whinchats, are welcome additions to the autumn mix and, depending on what is happening in continental Europe, thrushes will appear in greater or smaller numbers.

Big arrivals of passerines are rare and tend to be pale imitations of those on the east coast of England, for example. Two days after the arrival of hundreds of thousands of thrushes and finches on the east coast of England in late October 2012, Cape Clear was gently touched by the outer ripples of this movement: about 850 thrushes, including at least 12 Ring Ouzels. Sometimes, however, the island scores more heavily; one day in October 2005 birders found 33 Firecrests and 11 Yellow-browed Warblers, and a fortnight later there was a fall of 1,200 Robins. The island attracts a few autumn Hobbies (rare in Ireland), and Richard's Pipits and Common Rosefinches are semi-regular. And, of course, there is often an American vagrant awaiting discovery …

Working the island

Although the rewards can be great, Cape Clear does require effort; it is not a 'drive, park and start birding' destination. Planning is required. For a start, it is in the far south-western corner of Ireland where the roads are not fast, and there is a 40-minute boat trip from Baltimore to the island's North Harbour. A better strategy is to plan for a few days' stay and book accommodation, either at the observatory, the youth hostel or in one of the bed and breakfasts. That said, it is worth the effort.

Cape Clear is a windswept place and there are not that many decent-sized trees, but there is plenty of cover for migrants: leafy gardens, hedgerows, patches of gorse and heather and small copses. Some of the gardens are especially attractive to migrants.

Cotter's Garden and The Waist

Since 1962 the observatory, now operated by BirdWatch Ireland, has been close to North Harbour, a couple of minutes from the ferry berth and close to shops and bars. This is the hub of the island's birding and ringing programmes. Just to the south, sheltered Cotter's Garden is the stuff of legends, forever inscribed in birding history as the site of the Western Palearctic's first Blue-winged Warbler, which was found on 4 October 2000 in the aftermath of Hurricane Isaac. Being a mature garden, its rarity-attracting powers are immense. The first Irish Yellow-bellied Sapsucker took refuge in the garden in October 1988, spending most of the next few days clinging to the side of a small birch. In autumn 2005 Cotter's attracted Western Bonelli's, Greenish and Yellow-rumped Warblers in the space of six weeks. Nearby, the Post Office has pines outside, a magnet for any visiting Siskins or Common Crossbills. These are also popular with Redwings and Fieldfares in autumn. In a nearby garden, someone trying to re-find a Swainson's Thrush saw an unfamiliar bird appear where the thrush had been seen – it turned out to be a Pallas's Grasshopper Warbler. The small beach below The Waist hosted Ireland's first Black-and-white Warbler in 1978; in 2005 a Yellow-rumped Warbler fed among the seaweed on exactly the same stretch of beach. The Youth Hostel Garden has been a favoured refuge for Pallas's and Yellow-browed Warblers, Red-breasted Flycatchers and Common Rosefinches over the years.

East and West bogs, the new copse at Comar ('Thrush Glen'), Lough Errul and the gardens in Lighthouse Road are other places that are worth checking in spring, and even more so in autumn. When there are birds, a day is not enough time to explore everywhere.

Blannan headland

At the southern tip of the island is the legendary seawatching headland of Blannan, about half-an-hour's walk from the obs. This craggy spot has produced some amazing rarities, such as Ireland's only Bulwer's Petrel on 3 August 1975, Black-browed Albatross, several Macaronesian Shearwaters and a frigatebird species in 1973. Although easier-to-access sites on the mainland have overshadowed Blannan in recent years, it still produces the goods, particularly between July and September. Manx Shearwaters sometimes pass in thousands per hour in late summer, and it is an excellent vantage point to watch for Sooty, Cory's and (less so) Great Shearwaters, European Storm-petrels and skuas in late July and August. As if to demonstrate that seawatching can still be good here, three Wilson's Storm-petrels were seen one day in July 2011 and a Fea's-type petrel was found in mid-August 2012.

Blannan is usually best in the most unpleasant conditions: a south-west gale with poor visibility offshore. However, that is not always the case, as Steve Wing discovered on 11 September 2000. Then, in pleasant conditions and just a light breeze, he witnessed the passage of more than 5,000 Great Shearwaters in an hour, an event that left him totally perplexed.

Ireland's south coast (Tacumshin, Co. Wexford, in particular) is *the* place to see Buff-breasted Sandpiper (*Tryngites subruficollis*) in the British Isles, and the species sometimes turns up at Cape Clear in September.

- **LOCATION**
  Southern tip of Spain

- **FLYWAWAY**
  East Atlantic

- **SPRING**
  The season begins early with White Storks, a few raptors, Pallid Swifts and some passerines on the move in February; the first raptor peak is in mid-March, by which time seabirds, hirundines and many passerines can be seen; April is a busy month, and the first week of May is a second raptor peak, with large numbers of Griffon Vultures and European Honey-buzzards.

- **AUTUMN**
  White Storks and Black Kites peak in early August, followed European Honey-buzzards late in the month; other raptors – and most passerines – reach their peak flow in September and October.

- **KEY SITES**
  Cazalla, Guadalmesi, Punta Carnero, El Algarrobo and Puerto de Bolonia are all good watchpoints, depending on conditions.

- **THREATS**
  Wind turbines and overhead power lines are a threat to birds of prey; coastal development.

Tarifa, Spain

With just 16 kilometres separating the two blocks of land, and mountains on both sides, birds of prey can see Tarifa from Africa and vice versa. It comes as no surprise, then, that they cannot resist the temptation to use this route in and out of Iberia. For birders, the results are spectacular.

The Tarifa area, in the very south of Spain, offers the very best of European birding. It is actually a double bottleneck – raptor movements between Iberia and Morocco are concentrated here because the Strait of Gibraltar is so narrow. And seabirds passing between the Mediterranean and the Atlantic have no choice but to fly past it. And one must not forget the passerines; although not concentrated in the same way, the coast east and west of Tarifa is an excellent place to make contact with new arrivals from Africa in spring.

First and foremost, Tarifa is famed for its birds of prey and storks. Spring is great and autumn even better, though the definition of these seasons is somewhat loose. White Storks begin to cross into Spain in late January and, at the other end of the year, Merlins are sometimes still southbound in early December. That said, the busiest slots are from February to the end of May and from August to November. Different species peak at different times. In addition to the enticingly narrow crossing, the presence of high ridges on the Spanish side gives an additional boost to broad-winged raptors and storks because it enhances the formation of thermals, so important for their migration.

An early start to spring

Not all migration at Tarifa is visible, of course. Many migrants fly at night and

Tarifa is a major route in and out of Europe for Montagu's Harriers (*Circus pygargus*), which winter in Africa. The best time to see them here is in early April.

The old stone watchtower at Guadalmesi, looking across to the mountains of Morocco. This is pretty much the view raptors have in autumn as they prepare for the short sea crossing.

appear near the coast as if by magic. Millions more are too small to see with the naked eye as they pass overhead to reach their breeding sites further north. However, some pitch down in coastal scrub and on mountainsides, storks and raptors drift very visibly north and seabirds pass east and west through the Strait of Gibraltar.

Both migration seasons are prolonged affairs. Balearic Shearwaters pass east into the Mediterranean Sea between January and April, White Storks are on the move from late January and the peak passage of Great Spotted Cuckoo, Stonechat and Zitting Cisticola is in February. Cory's Shearwaters and Mediterranean Gulls migrate east from February to April and Little Gulls do likewise in March and April. At this time, Northern Gannets, Arctic and Great Skuas, Razorbills and Puffins are going the other way.

Early March is the peak period for White Stork, Black Kite, Short-toed Eagle, Thekla Lark, Meadow Pipit and Black Redstart. As the month progresses, Black Stork, Red Kite, Egyptian Vulture, Booted Eagle, Western

A light morph Booted Eagle (*Aquila pennata*). This African-wintering species peaks in late March and September and October in the Tarifa area.

Marsh and Hen Harriers, Osprey, Common and Lesser Kestrels reach their maximum spring numbers; a few Merlins also pass at this time. The first big raptor peak of spring comes in mid-March, with lots of species involved. March is also the busiest month for the arrival of Pallid and Alpine Swifts, Hoopoe, Eurasian Wryneck, Northern Wheatear, Ring Ouzel, Subalpine Warbler and Iberian Chiffchaff.

Spring gets hotter in more ways than one in April. Good numbers of the raptors moving in March are still on the move; now is the time that Montagu's Harrier, European Sparrowhawk, Common Buzzard and Eurasian Hobby reach their spring zenith. A very few Eleonora's Falcons are also involved. Non-raptors during April include Turtle Dove, European Bee-eater, Greater Short-toed Lark, Barn Swallow, Black-eared Wheatear, Rufous-tailed Rock Thrush, Western Olivaceous, Spectacled, Western Orphean, Garden, Western Bonelli's and Willow Warblers, Meadow, Tawny and Tree Pipits, Yellow Wagtail, Rufous Bush Robin, Common Nightingale, Bluethroat, Common Redstart, Whinchat, Pied Flycatcher, Woodchat Shrike, assorted finches and Ortolan Bunting. While Griffon Vulture passage is very prolonged, running from February to June, the first week of May usually provides the climax as it does for European Honey-buzzard. Late passerines, including Melodious Warbler, Spotted Flycatcher and Goldfinch, may still be on the move in June.

Autumn passage

During the autumn period passerines and near-passerines pass through the area in huge numbers but millions travel by night and even the day-flying migrants may be unnoticed apart from their calls high overhead. Some that are more obvious include Common Swift (from the second half of July), hirundines (mostly September and October) and finches (September to November). It is quite possible to find double-digit numbers of passerines such as Pied Flycatcher in relatively small areas. However, autumn is very much the season for migrant raptors and storks and, to a lesser extent, seabirds, and it is these above all that make the area special.

White Storks and Black Kites start to flood back to Africa in late July before peaking in early August. These two species dominate the large-bird migration until late in the month; with possible day-counts of more than 10,000 and 8,000 respectively. Seawatching can be productive, with Great and Arctic Skuas passing from the Atlantic Ocean to the Mediterranean from July, and Mediterranean Gulls passing in the opposite direction. Balearic and Yelkouan Shearwaters and Lesser Crested Tern can be seen at this time of year, and small numbers of Wilson's Storm-petrels have been noted on boat trips in the Strait in recent summers.

Also at this time, the raptor 'mix' changes dramatically, with European Honey-buzzards pouring south in their thousands (more than 10,000 are logged from a single watchpoint on a regular basis) in a very concentrated passage during the last week of August and the first week of September.

Montagu's Harrier numbers also increase at this time. Next, it's the turn of Short-toed and Booted Eagles, European Sparrowhawk, Osprey and Lesser Kestrel. Real rarities – the *cirtensis* subspecies of Long-legged Buzzard, Rüppell's Vulture and Eleonora's Falcon, for example – are regularly picked out of the kettles or loose flocks of commoner species. Long-legged Buzzard seems to have established a small breeding population in southern Andalucia; will Rüppell's Vulture follow suit? Later September witnesses the peak passage of Black Stork, Egyptian Vulture, Short-toed and Booted Eagles, Western Marsh Harrier, Goshawk, European Sparrowhawk, Osprey, Common and Lesser Kestrels and Eurasian Hobby.

Rüppell's Vulture

Rüppell's Vulture (*Gyps rueppellii*) is now categorised as Endangered due to a crash in the population across most of its range in Africa. The decline is attributed to habitat loss, the decline in wild ungulate numbers and persecution. One small cause for optimism is the increased frequency of sightings, since the 1990s, of birds crossing the Strait of Gibraltar with Griffon Vultures in spring and autumn. Those involved have mostly been sub-adults but it is hoped that a breeding population will be established in southern Spain or Portugal.

The great thing about Tarifa is that autumn just keeps on running: Red Kite, Hen Harrier and Common Buzzard peak in October, and Griffon Vulture and Merlin passage does not attain its greatest level until November, when some Red Kites, Hen Harriers, European Sparrowhawks and Common Buzzards are still on the move.

Top sites to visit

There are plenty of places from where to watch visible migration. For passerines this will occur on a broad front along the coast, but for the broad-winged birds of prey and storks it is important to check the weather forecast before deciding which to visit. Whether or not broad-winged raptors cross the straits depends on weather conditions. In spring, south-westerlies are good for encouraging their northward flights from the Moroccan coast. In autumn, after strong or prolonged easterlies they are more likely to 'launch' from the Puerto de Bolonia area, west of Tarifa; after prolonged westerlies they are more likely to set off around Punta Carnero, towards Algeciras. With calm conditions or just light breezes they congregate around Tarifa itself. Falcons, being much more active fliers, are less fussed by the niceties of the weather. On days when it is in progress, raptor migration is heaviest between 10.00 and 15.00 hours.

Situated high above Tarifa town, Cazalla is the classic watchpoint, where huge flocks of White Storks, Griffon Vultures, Black Kites and European Honey-buzzards can be seen at appropriate times, as well as Short-toed and

Red Kites (*Milvus milvus*) are year-round residents in much of Spain, but some leave the interior to winter in Morocco, passing through Tarifa from September to November and returning in late March.

Booted Eagles, accipiters and scarcer species. The intense coverage provided by Fundacion Migres raptor-counters ensures rarities are often picked out, including Eleonora's Falcon and regular Rüppell's Vultures. There is an information centre here. A little to the east, lower and closer to the coast, Guadalmesi, with its old watchtower, has great views across to Africa. Presumably, these are appreciated as much by the birds of prey as they are by human visitors. Scrubby areas are good for chats, thrushes and *Sylvia* warblers, as is the track that leads west along the coast towards Tarifa itself.

East again, just south of Algeciras and facing Gibraltar across the bay of the same name, is the headland of Punta Carnero. It is generally more productive in spring, when seabird passage includes shearwaters, Northern Gannet, Great Skua and Audouin's Gull. In strong westerlies, raptors may come in low. Inland is El Algarrobo, which can score heavily for raptors in autumn in westerlies. Rüppell's Vulture is regular here.

With a degree of good fortune, Rufus Bush Robins (*Cercotrichas galactotes*) can be found in coastal scrub when they return to Spain in May.

North of Algeciras, Los Barrios rubbish tip is on a raptor route and has a plentiful supply of rotting foodstuff for hungry Black Kites and vultures. It is not the most pleasant of places, but who cares when there's a chance of finding a Rüppell's Vulture? Eighteen species of raptors have been recorded in two hours, including Lanner, during spring migration.

If autumn raptors are the quarry but the weather is not good for migration, the Ojen Valley, about 30 kilometres north of Tarifa has been described as a 'waiting room', where they congregate until conditions improve. This is another regular site for Rüppell's Vulture and all the area's regular species are possible here, together with Pallid, Alpine and White-rumped Swifts. Passerine migrants can turn up in any patch of cover, and every post or overhead wire is worth checking for European Bee-eaters or Rollers. West again, La Janda is probably not as good as it once was but is still an excellent location for herons, raptors, waders and passerines. A winter visit may produce Black-winged Kite, Bonelli's Eagle, Common Crane, Little Bustard, Bluethroat and Moustached Warbler. Large numbers of Black Kite and Montagu's Harrier hunt here during migration periods, and the site is also good for passerine migrants.

MESSINA

Tyrrhenian Sea

CALABRIA

ITALY

mountains

Strait of Messina

SICILY

Reggio Calabria

- **LOCATION**
  Sicily, Italy

- **FLYWAY**
  Mediterranean–Black Sea

- **SPRING**
  Raptors, including rarities such as Amur Falcon, Black Vulture and Levant Sparrowhawk; storks; waders; passerines.

- **AUTUMN**
  Raptor passage lighter than during spring; storks, waders, passerines.

- **KEY SITES**
  Dinnammare and other raptor watchpoints; Capo Peloro lagoon for waders.

- **THREATS**
  Hunting, habitat destruction and large infrastructure projects such as power lines and construction of a bridge over the Strait of Messina.

Messina, Italy

In spring, the Strait of Messina raptor crossing is like a who's who of Western Palearctic birds of prey. No other site in Europe can claim such an amazing variety of hawks, harriers, eagles and falcons.

Messina is one of Europe's major raptor bottlenecks, particularly from March to May. As birds fly back towards their breeding sites in spring, they have already handled the long flight over the Sahara Desert and the Sicilian Channel (between Cap Bon, Tunisia, and the Italian island), which is 159 kilometres wide at the narrowest point. If weather conditions are not ideal – particularly if the birds are moving into a headwind, or there is a tailwind – they will head for the shortest crossing to the Italian mainland, the narrow Strait of Messina. There, there is a new and deadly hazard: poachers. Thanks to the efforts of conservationists over many years, this practice was outlawed on the Sicilian side of the Strait and is under control on the mainland (Calabria) side. In spring 2009, 40,000 raptors were counted in Sicily, and – for the first time in 26 years of the international effort – there were no gunshots. However, raptor hunting has since reared its head again. The Strait of Messina separates Sicily from the Calabria region of southern Italy. At its narrowest point it is only 3 kilometres wide, so very attractive for migrating birds of prey. The Peloritani mountain ridge runs almost west–east and it is from here that the best raptor watching, and counting, takes place.

Spring passage

More birds of prey use the crossing in spring than in autumn, and the most productive period is between mid-March and May. During these months in 2011, for example, 42,600 birds of prey and storks were counted. Within this time slot the most numerous species is European Honey-buzzard (with 35,736 in spring 2011), followed by Western Marsh Harrier (more than 3,414 one spring), Black Kite (more than 1,000

A Levant Sparrowhawk (*Accipiter brevipes*) photographed over Dinnammare, near Messina, on 10 May 2012.

Surely the most dramatic migration watchpoint in Europe? European Honey-buzzards (*Pernis apivorus*) spiral on a thermal as they prepare to cross the Strait of Messina, with the steaming hulk of Mount Etna providing the backdrop.

one year), Common Kestrel (almost 1,000) and Montagu's Harrier (866). In spring, Messina is the most important European flyway for European Honey-buzzards and the busiest in the Western Palearctic for Western Marsh and Pallid Harriers, Lesser and Common Kestrels and Eurasian Hobby. That much is predictable, but actual numbers are not. In spring 1992 more than 6,800 Red-footed Falcons were counted but in 2011 the figure was 'only' a still impressive 108. The list of rarer species is mouth-watering and includes Amur and Eleonora's Falcons, Black and Egyptian Vultures, Greater and Lesser Spotted, Eastern Imperial and Steppe Eagles, Levant Sparrowhawk, and Long-legged and Steppe (the *vulpinus* form of Common) Buzzards. In total, 40 species of raptors have been noted, a figure that cannot be matched anywhere in Europe. The crossing is also of international importance for White and Black Storks.

Timing

Passage is usually underway in mid- or late March, with Western Marsh, Hen and Pallid Harriers the first to pass over in reasonable numbers, accompanied by Lesser and Common Kestrels. During the first half of April, Western Marsh

Harrier is the most common species, with more than 1,000 possible in a single day. The normal running order for this species is adult males, adult females and second calendar year birds. Mid-April witnesses peak numbers of Black Kites; later in the month Montagu's Harriers are at their most frequent; and European Honey-buzzards only start to pass in good numbers in the last week of April. They, and Red-footed Falcons, peak at the end of April or in early May. Eleonora's Falcons do not reach their highest frequency until early to mid-May. Lesser and Common Kestrels, Black Kites and harriers can be seen at any time during April and May. Much depends on weather conditions. If it is sunny and warm, birds of prey find it easier to get a lift from thermals, so may cross higher and further south, where the Strait is wider. In wet and misty conditions, they may still cross but are more likely to do so at lower altitudes, and where the Strait is at its narrowest, just north of the town of Messina. They are most likely to be concentrated near Messina when the prevailing wind is from a northerly quarter in spring. Birds sometimes pass very low.

Migration in the Messina area is not just about raptors. In spring 2006, nocturnal passerines were tracked by radar as they passed over the Strait. In less than a month an estimated 4.3 million birds were tracked. In total, 327 species have been recorded.

In spring, Messina is the busiest flyway for European Honey-buzzards (*Pernis apivorus*) in Europe. More than 35,000 can pass through in a single season.

Working the site

Passage can start early so it is worth doing as the raptor-counters do and putting in a full shift, starting at 7 a.m. Dinnammare is the highest observation point, 1,127 metres above Messina. As visible migration watchpoints go, this is one of the most dramatic in the world, with Mount Etna providing an awesome backdrop. It is best when the wind is blowing from the north-west, but even if the harriers, buzzards and falcons are in short supply it is worth being there for the view. The maritime climate of the area greatly influences the raptors' movements and where they can be viewed. If low cloud suddenly descends or the wind switches, be prepared to move elsewhere. There are plenty of sites to choose from: Gesso, Locanda, Forte Ferraro, Chiarino, Puntale Bandiera, Portella, Castanea and Monache, the last being the closest to Messina town.

Capo Peloro

If viewing conditions are not optimal, there are other options. Capo Peloro lagoon, a few kilometres north-east of Messina and pretty much the easternmost point in Sicily, is great in spring and autumn for waders and herons: look for Spoonbill, Glossy Ibis, Black-winged Stilt, Marsh Sandpiper, Temminck's Stint, Slender-billed Gull and Black, White-winged Black and Whiskered Terns. With an onshore wind nearby Capo Peloro headland can be good for seawatching, including Scopoli's and Yelkouan Shearwaters, Audouin's and Little Gulls, and Gull-billed, Caspian, Black and Common Terns. In southerly winds the Torre Faro raptor viewpoint here sometimes comes into its own. From here it is possible to see thousands of small birds migrating during the day: Common and Pallid Swifts, Skylark, swallows and martins, wagtails, Golden Oriole and finches. Sometimes, visible migration continues all day. Migrant Eurasian Wryneck, Red-rumped Swallow, European Bee-eater, Tawny Pipit, Black-eared and Northern Wheatears, Rufous-tailed Rock Thrush, several species of *Sylvia* warblers and flycatchers can be found. Rarer species such as Great Spotted Cuckoo are not unknown.

Eleonora's Falcon (*Falco eleonorae*) nests on cliff ledges on offshore islands. It returns to Sicily in late April and early May.

HUNGARY

Egyek

Halastó
fishponds

HORTOBAGY
NATIONAL PARK   Hortobágy

pusztas

pusztas

- **LOCATION**
  Eastern Hungary

- **FLYWAY**
  Mediterranean–Black Sea

- **SPRING**
  Waders; marsh terns;
  passerines, including
  Moustached Warbler.

- **AUTUMN**
  Wildfowl, including Lesser
  White-fronted and Red-
  breasted Geese and Smew;
  Common Cranes; raptors
  may include Pallid Harrier;
  waders, including Dotterel;
  passerines.

- **KEY SITES**
  Halastó fishponds, Darassa
  and Egyek pusztas, Ferek-
  ret, Lake Tisza.

- **THREATS**
  Protected as a national
  park.

Hortobágy, Hungary

*Line after line of trumpeting Common Cranes, washed
pink by the sunset, approach and descend to join a roosting
throng in a shallow fishpond. This is one of the great
sights of Hortobágy National Park in autumn.*

Since the start of the new millennium the cranes' eastern European flyway
has brought much-increased numbers to Hungary's Great Plain each
autumn. It is these birds that have become the Hortobágy's main claim to
fame. From late September into October tens of thousands of cranes use
the area as a staging post on their long migration. While they are in the
Hortobágy the cranes forage by day on the *puszta* (steppe) and roost by night
on fishponds. They arrive from breeding areas in Finland, the Baltic States,
northern Poland and Belarus. When they depart, they do not all head for the
same destination. Some winter relatively close in Serbia, for example; others
end up in Tunisia, Ethiopia or Eritrea. Satellite-tagging has provided much-
needed information on cranes' migration patterns. For example, in autumn
2011 the satellite-tagged crane Horsma left Finland in mid-September and
arrived in the Hortobágy on 5 October, where it remained until 10 November.
Rather than continuing to Africa, it divided the winter between southern
Hungary and Serbia. The following spring, this bird stopped over for just a few
days in the Hortobágy before continuing north.
A much shorter spring stopover is typical.
Situated on the Mediterranean–Black Sea
Flyway, eastern Hungary has much to offer in
spring and autumn. Tens of millions of birds
use this route between central and eastern

**Common Crane (*Grus grus*) is
the flagship autumn migrant
in the Hortobágy. Thousands
descend on the *pusztas* and
fishponds from late September.**

**This wetland near
Lake Tisza is typical
of much of the area,
with shallow pools,
emergent vegetation,
muddy fringes – all
surrounded by *puszta*
grassland.**

Europe and western Siberia to the north and the Middle East and Africa.
Countless birds – unseen nocturnal travellers – fly straight over the Hortobágy,
particularly if weather conditions are favourable. But for those that do not,
the area offers plentiful food and relative shelter. It is Hungary's largest area
of *puszta*, dominated by grasslands and arable farmland. Depressions contain
pools and marshes, with seasonal wetlands (for example, Fekete-ret) appearing
in spring and autumn. Hundreds of fishponds, such as those at Halastó, cover
an area of about 5,000 hectares. Some of these have reedy fringes or muddy
margins. All these wet areas are breeding grounds for billions of invertebrates
so are very attractive to a wide range of breeding and migrating birds.

Spring

In late March the first returning White Storks arrive at their traditional nest sites in Hortobágy villages, such as Nagyiván. Halastó is a spring stopover site for a few Lesser White-fronted Geese that punctuate their transit from Greece to Russia and Scandinavia and feed with the commoner wintering geese for a while. Wildfowl numbers increase as birds that have wintered further south stop off en route to northern latitudes. Parties of Northern Lapwings and other waders rest and feed on damp grassland and by pools on their way to Poland, Belarus or the Arctic. The first Lesser Spotted Eagles and Pallid Harriers of the year appear, bound for breeding territories to the north-east. Early passerine migrants such as Black Redstarts pass through the area, some remaining to breed in local villages where they are very common.

April and May witness the passage of thousands of wildfowl and waders. Garganey return to breed. Red-footed Falcons and a few Montagu's Harriers return to the *puszta*. Little Gulls, marsh terns, swallows and other passerines reappear. If the weather is fine, breeding birds – including Moustached Warblers and hundreds of Sedge, Reed, Great Reed and Savi's Warblers – suddenly reappear on territories they vacated the previous autumn. If migration is held up by bad weather many more small birds will be obvious as they are forced to break their northward journeys for a while. The whole eastern European cast will be represented, from Red-throated Pipits to European Rollers, Barred Warblers to Red-backed Shrikes, and Great Reed Warblers to Golden Orioles.

Western Marsh Harriers (*Circus aeruginosus*) are common breeders but depart the Hortobágy in autumn before the pools freeze. This raptor is also a passage species, stopping over to and from its enormous eastern European breeding range.

Autumn return passage

In August the muddy margins of the fishponds attract returning waders. Numbers build up in September, particularly on pools that have been drained. Big numbers of Ruff, Black-tailed Godwit, Curlew Sandpiper, Little Stint and Eurasian Curlew can be seen, with scarcer species including Broad-billed and Marsh Sandpipers, Ruddy Turnstone and Red-necked Phalarope. In September 2007 Hungary's eighth Buff-breasted Sandpiper was at Halastó. One of the real specialities of the Hortobágy grasslands is Dotterel. The first trips arrive in late August and numbers increase to a peak a month later – more than 200 in 2007. In October they continue on their way south. During September large numbers of swallows and other passerines move through. Passage Red-footed Falcons join birds that have summered in the area, and a few migrant Pallid Harriers are seen.

More and more wildfowl appear on the pools and fishponds, with large numbers of passage Northern Shoveler and Eurasian Teal joining breeding species such as Ferruginous Duck before moving south again. Tens of thousands of geese – mostly Greater White-fronted, but also Tundra and Taiga Bean Geese and Greylag Geese – also appear. They feed on arable land by day and roost on the shallow lakes at night. As autumn turns to winter Red-breasted and Lesser White-fronted Geese will be found among them. The former have increased in recent years, but numbers of the latter species have fallen. Dozens of Smew arrive to spend the winter on the fishponds. Merlins from the north (and resident Sakers) hunt over the grasslands and White-tailed Eagles hunt over the wetlands. Late autumn rarities have been found: a Pine Bunting was trapped in November 2011. Pride of place, though, goes to the spectacular visitation by Common Cranes. The first arrive early in September and by the end of the month in 2011 there were 60,000, with numbers peaking at 100,000. With the onset of the first harsh frosts they continue on their migration south.

Working the sites

The best area of fishponds to explore are the 'Great Fishponds' at Halastó, just west of the village of Hortobágy. Halastó has a visitor centre with notification of recent sightings and a narrow-gauge railway that rattles along not much faster than walking pace to the biggest fishpond in this group, Kondás. On the way it passes between several large fishponds and some large chunks of reedbed, occasionally broken by stands of willow. However, it is usually more productive to walk alongside the railway tracks to the end, a distance of 4.5 kilometres, checking each pool along the way. Near the start of the tracks, a series of small reed-fringed pools separated by dykes is worth checking for crakes, waders, Reed, Sedge

A reeling male River Warbler (*Locustella fluviatilis*). This denizen of riverside vegetation and swamp margins is a late arrival, appearing mostly from mid-May. Eastern Hungary represents the southern-most part of its range.

and Great Reed Warblers, Penduline and Bearded Tits, Bluethroats and other passerines. In spring, Savi's Warblers reel from the more extensive reedbeds beyond. These can be viewed from a series of tower hides. There is also a small population of Moustached Warblers. Wildfowl, Pygmy and Great Cormorants, Little Bitterns, Grey, Purple, Squacco and Black-crowned Night Herons, Little and Great Egrets, Spoonbills, Glossy Ibis and Common Cuckoos fly from one fishpond to another. Ditches hold crakes.

In early spring, before their move north, thousands of ducks occupy the bigger ponds and many more geese feed in the fields beyond, moving to the fishponds at night. In autumn, a good selection of passerines turn up in cover along the main tracks; check in any clumps of willow. Near the end of the railway, the very tall tower hide gives extensive views over Kondás and fishpond V. Kondás is one of the cranes' favoured roost sites, and the tower hide is an ideal spot from which to scan them in September and October, when there are also hunting White-tailed Eagles and Western Marsh Harriers. To the east, fishpond VI is good for marsh terns, mainly Whiskered but with some White-winged Black and Black, particularly during passage. Halastó fishponds require a permit, which can be purchased on-site. This requirement is strictly enforced.

The muddy margins of drained fishponds at Halastó and elsewhere in the Hortobágy are perfect food sources for hungry migrant Wood Sandpipers (*Tringa glareola*).

Darassa and Egyek *pusztas*

Several areas of *puszta* are easily accessible by road. Just south of the road between Tiszacsege and Balmazujvaros, a tower hide overlooks the Darassa *puszta*, though it is now in a very rickety state. This is a good place to watch feeding cranes in September and October and also to scan for Sakers and Montagu's Harriers. South of Halastó, Fekete-rét can attract passing Dotterel in autumn, as well as large numbers of geese. The road running south from highway 33 to Nagyiván village is good for both passage and breeding Red-footed Falcon and other raptors, larks, pipits, wagtails, Northern Wheatear and Lesser Grey and Red-backed Shrikes. That said, any of the roads that cross the Hortobágy, particularly where they pass pools and damp depressions, are likely to be productive for these and other migrants such as White Stork, European Bee-eater and Roller. Small copses, planted as windbreaks, often hold Golden Orioles.

The largest of the marsh terns, Whiskered Tern (*Chlidonias hybrida*) arrives at fishpond breeding sites in late April. White-winged Black (*C. leucopterus*) and Black (*C. niger*) Terns also breed.

BULGARIA

CAPE KALIAKRA

Varna

Black Sea

BURGAS
LAKES

Burgas

- **LOCATION**

  Western flank of Black Sea, Bulgaria

- **FLYWAY**

  Mediterranean–Black Sea

- **SPRING**

  Waders from early March–May. White Storks mid-March to mid-April. Most raptors peak in April. Passerines and near-passerines in April–early May. European Honey-buzzards in late April–early May.

- **AUTUMN**

  Return passage starts in July, becoming stronger in August. White Storks and European Honey-buzzards peak in late August–early September. Passerines peak in September–early October, with raptors and White Pelicans in late September. Wildfowl on the move in October–December.

- **KEY SITES**

  Burgas wetlands: Atanasovsko Lake, Vaya Lake, Mandra Lake with Poda Lagoon; Pomorie Lake; Kableshkovo Hills; Cape Kaliakra; Shabla lakes; Durankulak Lake.

- **THREATS**

  Coastal leisure development, wind farms, illegal shooting.

Via Pontica, Bulgaria

The Via Pontica migration route, running along Bulgaria's east coast, has been known for many years, but it only appeared on most western birders' radars in the 1990s. This coast offers visible migration at its very best.

For birds vacating breeding areas in Russia, Belarus, Ukraine and the Baltic States to winter around the eastern Mediterranean or in Africa, the Black Sea presents a major obstacle. Some migrants cross it directly but millions of others fly either around the east – into eastern Turkey, Iraq and the Gulf States – or the west, through Romania and Bulgaria, before crossing the Bosphorus into Turkey. Three flyways have been identified running north–south through Bulgaria, the most important of which is the easternmost: the Via Pontica. This is the central arm of the great Mediterranean–Black Sea flyway, one of three Palearctic–African migration conveyor belts that together transport 2.5 million geese and ducks, 2 million raptors and 2 billion passerines and near-passerines between breeding grounds in central and eastern Europe and western Siberia, and winter quarters in tropical and subtropical Africa.

Cape Kaliakra protrudes into the Black Sea not far south of the Romanian border. It is a great site for visible migration (including raptors) and can also be good for seawatching.

Large saltpans, reed-fringed freshwater lakes and estuaries along the Black Sea coast provide pit-stops for tens of thousands of migrating waders. Just inland, coastal ridges create updrafts and thermals for soaring birds (pelicans, storks and birds of prey) to gain height. Up to 734,000 of these use the flyway in spring and up to 922,000 in autumn. Headlands such as Cape Kaliakra thrust out into the Black Sea, providing welcome resting places for tired passerines that have crossed the sea, bound for the vast plains and forests of Russia. In autumn Cape Kaliakra is also a jumping-off point for southbound birds, large and small. Other small birds, among them incalculable numbers of hirundines, track along the coast.

A male Montagu's Harrier (*Circus pygargus*). Sadly, the species has declined as a breeder, but good numbers pass south in autumn from populations farther north.

In addition this magical stretch of coast has populations of some very special breeding birds, including Pied Wheatear, Paddyfield Warbler, Semi-collared Flycatcher and Black-headed Bunting. So without ever leaving the flyway, a fortnight in late April–early May or late September, divided between the wetlands around Burgas and the coast and wetlands near Kavarna, will produce around 200 species – and some dramatic visible migration.

White Storks lead the way

Spring starts early. White Storks start to appear early in March, their northward migration climaxing in early April. They are the most numerous of the large soaring migrants both in spring and autumn. Early waders such as Dunlin, Northern Lapwing, Kentish Plover, Black-winged Stilt, Pied Avocet, Black-tailed Godwit, Ruff, Common Redshank, Spotted Redshank, Common Greenshank and Wood Sandpiper appear at the wetlands around Burgas and further north. In April Black-winged Stilt, Eurasian Curlew, Marsh, Common and Curlew Sandpipers, Little Stint and Common Ringed Plover are added to the feeding and roosting flocks. The most numerous species are Ruff, Curlew Sandpiper, Pied Avocet, Little Stint, Northern Lapwing, Common Redshank and Spotted Redshank. Scarcer birds such as Temminck's Stint, Broad-billed and Terek Sandpipers and Sanderling may require a little searching to find.

The most productive wader sites are Pomorie and Atanasovsko Lakes. Great White Pelicans, Glossy Ibis, Purple Herons and raptors are on the move throughout April, with large flocks of Lesser Spotted Eagles, Common and Steppe (*vulpinus*) Buzzards, and Red-footed Falcons possible from mid-month. The last raptors to pass through are European Honey-buzzards and Levant Sparrowhawks, which track north mostly in mid- or late May. April and May are the prime months for passerine and near-passerine migration. In early May Common Cuckoos, Turtle Doves, Nightingales, Golden Orioles, Barred Warblers and Red-backed Shrikes are in any patch of suitable cover. European Bee-eaters and Rollers perch on wires. Everything seems to be on the move.

Raptor extravaganza

July sees passage waders returning to traditional sites and these birds' numbers build during August. Then, White Storks leave breeding grounds in Ukraine, Belarus, the Baltic States and Poland and aggregate in spectacular flights south along the flyway. Late August is a great time to see them in coastal Bulgaria en route to the Bosphorus, Eilat and – ultimately – sub-Saharan Africa. When conditions are right it is not unknown for observers to count 70,000 or more passing overhead in a single day. Adult European Honey-buzzards are also in full flow in late August and early September, with young birds following later than their parents. September is also a time when passerines make short movements along the coast, feeding up at headlands before their proper migration starts. As the month progresses, the stream of White Storks dries to a trickle but there is an absolute torrent of White Pelicans, Black Storks, Lesser

Semi-collared Flycatcher

BirdLife considers this declining species to be Near Threatened. It has suffered from deforestation within its breeding range, but not enough is known about its wintering requirements to determine the problems it faces there. It breeds in Bulgaria, Greece, Turkey, Georgia, Azerbaijan, the extreme south of Russia and north-east Iran, and winters in a relatively small area of East Africa from Sudan and South Sudan through western Kenya, eastern D.R. Congo, Uganda, Rwanda and Burundi to Tanzania. The European population is estimated to be in the range 15,000–53,000 pairs. Of this, the combined Bulgarian and Greek population is thought to be just 2,500–6,500 pairs. In Bulgaria it is spread thinly across much of the hillier parts of the country, favouring deciduous and mixed woodland, generally at higher altitudes. Semi-collared Flycatcher is unlikely to be seen on passage but it does breed in parts of the Via Pontica flyway, including in Goritsa Forest, near the Kableshkovo Hills. Birds arrive on breeding territories from mid-April and start their return passage in late July.

Spotted Eagles and other raptors, hirundines (exceptionally, 100,000 Barn Swallows can be seen in a day) and other assorted passerines. There is then a lull before the late-autumn influx of wildfowl, moving south in advance of the winter freeze-up further north.

A sumptuous pink-flushed male Red-backed Shrike (*Lanius collurio*). This species is a common breeder and passage migrant in spring and autumn, though not as common as it once was.

The Burgas area

Around the city of Burgas, several large lakes and wetlands demand to be explored. In spring and autumn they host migrant wildfowl, pelicans, waders, gulls and terns. In winter their appeal is very different, when they pull in thousands of geese and ducks (including hundreds of Greater White-fronted and some Red-breasted Geese and hundreds of White-headed Ducks) and thousands of Pygmy Cormorants.

Just north of the city, Atanasovsko Lake is shallow and salty, part of a complex of saltpans that connects with the Black Sea. There are also marshes, reedbeds and freshwater pools. More than 300 species have been recorded at this Ramsar site, a large proportion of them migrants. Sadly, the most famous

A forest-breeding species of the Bulgarian interior, Collared Flycatcher (*Ficedula albicollis*) does not breed along the coast. It may, however, turn up on the Via Pontica on passage.

of them all – Slender-billed Curlew – is now one of the world's rarest birds and could be extinct. It was last reliably seen at Atanasovsko Lake in 1993. An impressive array of waders, raptors, herons and egrets, Spoonbills, White and Black Storks, White and Dalmatian Pelicans, Little, Slender-billed and Mediterranean Gulls, Gull-billed and marsh terns pass through in spring and autumn. Others, such as Pied Avocets, Black-winged Stilts, Kentish Plovers, Collared Pratincoles, Sandwich and Little Terns and *Acrocephalus* warblers go no further north but stop to breed. Water Rails and Little and Spotted Crakes wander around in reedy ditches.

Just west of Burgas city centre, brackish Vaya Lake is the largest natural lake in Bulgaria and has the country's most extensive reedbeds. It is an important stopover location for migrant wildfowl, raptors and passerines. It does not freeze entirely in winter, so experiences influxes of grebes, Pygmy Cormorants, wildfowl (including White-headed Ducks) and Dalmatian Pelicans when harsh conditions freeze other lakes further north. A little to the south is the Mandra-Poda lake complex, with Poda Nature Reserve at the eastern end, adjacent to Burgas Bay. The lakes are a Ramsar-listed wetland, complete with reedbeds and muddy margins, hides and a visitor centre. There is an impressive mixed colony of herons and White-headed Ducks visit in winter.

Kableshkovo Hills

A few miles north of Burgas, the hills around Kableshkovo come into their own in autumn because this area seems to benefit from the most concentrated passage of soaring birds. It boasts average counts (2004–2011) of 150,000 in spring and 270,000 in autumn. The last week of September is often the best time and the movement can be spectacular, especially if weather has been poor over the previous few days. Then, raptors will have been forced to roost on forested Balkan slopes and at the first sign of a break in the weather will be keen to get going again. The first indication that something major is happening is a handful of distant specks that imperceptibly get bigger and morph into a spiralling kettle of 20, 30 or 40 birds. As they get closer, identification becomes possible.

On a good day a pulse of raptors and storks may last for three hours or more. For example, on 22 September 2007, between 9:00 a.m. and 11:00 a.m., the following were logged: at least 3,000 White Pelicans, 2,000 Lesser Spotted Eagles, 600 Common Buzzards and 150 Black Storks. Expect to pick out a few Short-toed Eagles, European Honey-buzzards and harriers, while making more direct progress could be Goshawks, Levant and Eurasian Sparrowhawks, Red-footed Falcons, Common Kestrels, Eurasian Hobbies and the odd Osprey. Twenty species of raptors in a day is possible.

Pomorie Lake
About 25 kilometres north-east of Burgas, and just north of the town whose name it bears, the saltpans here hold many waders in spring and autumn, as well as the *de rigeur* herons and egrets. In spring some massive flocks of Ruff have been seen here, along with smaller numbers

Purple Heron (*Ardea purpurea*) is a summer visitor, breeding in small numbers in coastal wetlands, especially those around Burgas. Migrants also track the coast to and from breeding grounds to the north.

of Curlew Sandpiper, Little Stint, Kentish Plover, Common Greenshank and Marsh Sandpiper. Scarcer species such as Red-necked Phalarope are possible. As with everywhere along the Bulgarian coast in spring and autumn, 'eyes to the skies' is a good rule of thumb since stork and raptor migration may be visible overhead. Interesting passerines are always likely.

Cape Kaliakra

This limestone headland, near Kavarna, spikes south several kilometres into the Black Sea. The cape towers nearly 100 metres above the water, making it a magnet for migrants crossing the sea from points east, and especially for those moving south along Bulgaria's east coast. It is good for passerine migrants in spring (including breeding Pied Wheatears) and Rosy Starlings often put in an appearance in early summer.

In autumn Cape Kaliakra is particularly good for raptors, though not in the numbers that are possible at Kableshkovo. Look for European Honey-buzzard, Lesser Spotted, Booted and Short-toed Eagles, Pallid and Montagu's Harriers, Common and Long-legged Buzzards, Red-footed Falcon, Goshawk, Eurasian Sparrowhawk and Levant Sparrowhawk, which are all possible. Its limestone grasslands provide a well-used stop-over for Quail and Corncrake. Eurasian Stone-curlew, Calandra, Greater Short-toed and Crested Larks, European Bee-eater, hirundines, pipits, wagtails, chats, warblers, Red-breasted Flycatcher, European Roller and Red-backed Shrike are all likely here or nearby, some in huge numbers. In late September more than 100,000 Barn Swallows have been counted in a single day in this area. Yelkouan Shearwater can be seen offshore.

A female Pied Wheatear (*Oenanthe pleschanka*). The region around Cape Kaliakra represents the western extremity of this species' range. Pied Wheatears return to the country in April but the breeding population is estimated at just 350–500 pairs.

Lake Durankulak

A little further north, close to the border with Romania, is Lake Durankulak. This freshwater lake has extensive reedbeds and is close to the Black Sea coast. Its position makes this lake particularly attractive for water birds. Thousands of wildfowl – notably large numbers of Red-breasted and Greater White-fronted Geese – spend the winter in the area. By February the geese graze almost exclusively in the cereal crop fields surrounding the lakes of Shabla and Durankulak, and roost on the lake. Goose movements can be complex. If the weather is mild, flocks may leave the main herds and fly north in late February, only to return again if harsh weather sets in further north. This can make for an exciting spectacle. At the end of February 2012, for example, there were 6,000 Greater White-fronts and 2,500 Red-breasted Geese in one field, feeding on harvested maize. The main spring exodus is in March and April. At this time, wildfowl migrating from the southern section of the Bulgarian Black Sea coast use the area as a staging post.

Glossy Ibis, Spoonbills, Little Bitterns, Levant Sparrowhawks, Red-footed Falcons, Little and Spotted Crakes, Collared Pratincoles, Gull-billed, Caspian, Whiskered and White-winged Black Terns, and commoner migrants pass through or take up residence for the summer in April and May. In autumn, thousands of water birds take over, including Black-necked Grebe, Pygmy Cormorant, Eurasian Wigeon, Common Pochard, Tufted Duck, Mediterranean Gull and White-tailed Eagle.

A drake Ruddy Shelduck (*Tadorna ferruginea*). Numbers of this summer visitor to Bulgaria are greatly reduced, but there is a small population on the lakes around Burgas.

- **LOCATION**

  In the north-east Aegean Sea, just off the west coast of Turkey

- **FLYWAY**

  Mediterranean–Black Sea

- **SPRING**

  Passage begins in late February with the first Northern Wheatears, then builds through March to a climax in the last week of April and first week of May. Wildfowl, herons, raptors, crakes, waders and passerines all feature.

- **AUTUMN**

  Waders, raptors and passerines, but autumn passage is a trickle compared with spring.

- **KEY SITES**

  Kalloni and Polichnitos saltpans, Skala Kallonis pool, Ipsilou, Metochi Lake, Faneromeni and Sigri fields, Meladia Valley, Kavaki, Napi Valley, Potamia Valley.

- **THREATS**

  Wind farm development.

Lesvos, Greece

Sitting just off the coast of Turkey, the Greek island of Lesvos is the classic east Mediterranean migration site. Interest is astonishingly varied but its shining lights are egrets, storks, crakes, waders and passerines in spring (and to a lesser extent in autumn). The island also boasts several breeding specialities, adding to its overall appeal.

Several factors help explain why Lesvos has such an impressive track record for migrant birds. Its north coast is just 8 kilometres from the coast of Turkey. Millions of birds heading to and from breeding ranges in eastern Europe and Asia Minor fly over this part of the Mediterranean every spring and autumn. Lesvos is a relatively large island (about 70 kilometres from west to east) and has a much more varied mix of habitats than any other Aegean island, indeed more than much of the Turkish coast – and that is why it is such a stand-alone location for spring migration. And there are plenty of birders covering the ground, especially in the core weeks of April and May. This observer coverage helps to explain the high proportion of rarities on the total Lesvos list of more than 320. An impressive selection of breeding specialities, including Ruddy Shelduck, Chukar, Rufous Bush Robin, Rüppell's Warbler, Krüpers and Western Rock Nuthatches, Masked Shrike and Cretzschmar's and Cinereous Buntings add flavour.

Lesvos has a long coastline, including saltpans that exert a magnetic attraction on thousands of herons, waders, gulls and terns. Storks, pelicans and birds of prey are also drawn to them. In spring, at least, it has vegetation-fringed freshwater pools where Garganey, crakes and Little Bitterns can lurk. Large open areas are ideal for larks, pipits and hunting raptors. And many of the valleys have plenty of cover for shrikes, warblers and flycatchers. The island is drier in autumn so loses much of its wetland appeal at that time. The few remaining bodies of fresh water become the focus for many species.

The year in focus

The first Northern Wheatears appear in late February, and northbound migration accelerates through March, when Kalloni saltpans and other suitable sites witness their first Black and White Storks, Glossy Ibis, Spoonbills and Great White Pelicans. At the end of the month any suitable wetland may hold Black-crowned Night Herons, Purple Herons and Little Crakes. In 2010 a Yellow-browed Warbler turned up in late March near the mouth of the Vergias River; this was the first record of the species for Lesvos and only the tenth for Greece.

The most exciting period begins in April. Sometimes the pace of change can be bewildering. Saltpans attract Garganey, herons and storks, along with hundreds of waders, gulls and terns. At freshwater pools there will be egrets, Little Bitterns, Purple Herons, crakes, marsh terns and wagtails (including Citrine and sometimes hundreds of Yellow). Larks, pipits, wheatears and buntings will appear in areas of open grassland and rocky hillsides, with newly arrived Red-footed Falcons, Lesser Kestrels and other raptors hunting overhead. Bushy country, fig and olive groves, and any other area with cover will hold a potpouri of chats, warblers, flycatchers, shrikes, Golden Orioles and buntings (Cinereous, Ortolan and Cretzschmar's arrive mid-month). Some of these species breed on the island; many more use it as a temporary stopover.

The seasonal wetlands at Alykes, looking east towards Mount Olympus. In spring these attract Garganey, herons, crakes and waders. Most Lesvos wetlands dry out over the course of the summer.

If weather conditions deteriorate while migration is underway, a stream of birds may become a deluge, as happened on 19–20 April 2011 when cold, wet and windy weather forced many – including tens of thousands of Collared Flycatchers and many *Sylvia* warblers, Nightingales, Turtle Doves, chats and even Semi-collared Flycatchers – to pitch down on Lesvos. This is the period when rarities are found: for example a Finsch's Wheatear in the second week of April 2010 and many Semi-collared Flycatchers in the previously mentioned fall of 2011. After only 19 records prior to 2011, there were then 33 records of 40 individuals, surely just a one-off? The island's first Steppe Eagle was in the last week of April 2008.

Some of the best birds arrive in May. Early in the month is a good time to look for Dalmatian Pelican at saltpans, though it is a rare bird on Lesvos. European Bee-eaters and shrikes often arrive in large numbers – virtually anywhere – during the first week. Although a much scarcer bird, European Roller is a possibility, while flocks of Rosy Starlings arrive from mid-month. Genuine rarities could include Blue-cheeked Bee-eater.

Coastal saltpans provide perfect feeding for stopover Curlew Sandpipers (*Calidris ferruginea*) as they migrate from Africa to the Arctic tundra in spring. After leaving Lesvos, they fly north, overland. The species is scarce in autumn.

Wetlands dry out

The island is much drier in autumn and this is reflected in fewer sightings of freshwater-loving migrants. This is true for species such as Little Bittern, Black-crowned Night Heron, Purple Heron and marsh terns, for example, and crakes and Collared Pratincole are virtually unknown in autumn. However, return wader passage begins in July when both Polichnitos and Kalloni saltpans have plenty on offer. The former is actually better than it is in spring.

Returning Garganey can show up from late July. Waders dominate in August, though late in the month migrant passerines are more in evidence. The biggest flock of Great White Pelicans (144) was seen in September (2007), a month in which rarities may occur; for example, in 2008 the island's only Steppe Grey Shrike. Wintering wildfowl slowly increase in September and October.

Saltpans and freshwater pools

No single location dominates and migrants can turn up pretty much anywhere on the island. At prime times it is worth checking any bit of habitat that looks as though it has potential – particularly if it has fresh water or some trees to provide cover. However, the default site to check is the Kalloni saltpans at the north end of the sheltered Gulf of Kalloni. It is a large site and cannot be worked quickly. In the winter months it is good for wildfowl and some waders, but come March it really starts to hot up and can be crammed with birds.

Between late March and May this is a good first port of call for Garganey, White and Black Storks, Glossy Ibis (occasionally more than 100), Spoonbill and – with a big dose of luck – Dalmatian Pelican. Thirty or more species of waders can be seen here in late April or early May, and a rarity is possible. A Broad-billed Sandpiper, Spur-winged Plover or Great Snipe is a possibility, and a Terek Sandpiper was recorded in 2011. The waders most likely to be encountered include Black-winged Stilt, Pied Avocet, Kentish Plover, Wood Sandpiper (sometimes more than 100), Little Stint (more than 300 in early May 2011), Curlew Sandpiper, Ruff (800 in spring 2012) and Collared Pratincole. The neighbouring grassland and seasonal pools of the Alykes wetlands are better in spring than autumn. They are worth checking for herons, marsh terns, pipits (large numbers of Red-throated are possible) and wagtails.

Skala Kallonis and Polichnitos saltpans

Just west of the saltpans is the small resort of Skala Kallonis. Its pool, while not as productive as it once was, is still worth checking if it holds water. A short walk west of Skala Kallonis, the Christou River flows into the Gulf of Kalloni with a small marsh adjacent. There is a viewing screen here, from where herons, storks and waders can be seen.

On the eastern edge of the Gulf of Kalloni, the Polichnitos saltpans are a smaller and faster-to-work alternative to Kalloni. The same mix of species is possible and the site retains its pulling power in

An adult male Woodchat Shrike (*Lanius senator*) of the nominate race. This is a common summer breeder and passage migrant in April and May and between August and October. Sometimes there are double-figure counts at favoured locations in spring.

autumn, which begins in July when the first Arctic-breeding waders return. A Pectoral Sandpiper here in September 2008 was the island's first. Saltpans, river estuaries and any other wet areas are worth checking for passage gulls, including Slender-billed, and terns. Whiskered is the commonest of the marsh terns and Black the scarcest; all three species, and Gull-billed Tern, are more likely to be seen in spring than autumn.

Crakes, raptors and passerines

For variety, Metochi Lake (on the Christou River west of Kalloni) is worth a visit. It is a prime spot for crakes, with Little the most likely and Baillon's distinctly unlikely – though still possible. A spring visit could produce a selection of herons, waders and marsh terns. Unlike many pools on Lesvos it rarely dries out completely in autumn so can still attract waders. Migrant chats and shrikes should be found between August and October.

To the north of Kalloni, on the road to Petra, is a bandstand that functions as the Kalloni Raptor Watchpoint in spring. In May 2009 and April 2011 single Eastern Imperial Eagles were seen here, and other Lesvos scarcities or rarities include Pallid Harrier, Levant Sparrowhawk, Lesser Spotted and Booted Eagles and Eurasian Hobby. More likely will be European Honey-buzzard, Short-toed Eagle, Western Marsh and Montagu's Harriers, Common and Long-legged Buzzards, Lesser and Common Kestrels, Red-footed Falcon

Citrine Wagtail (*Motacilla citreola*) occurs in small numbers most years, typically in the second half of April or the first week of May. This bird was at Faneromeni in May 2012.

and Eleonora's Falcon. In May 2003 one lucky observer counted 73 of the last species over in just one hour.

North of the village of Sigri, in the far west of Lesvos, Faneromeni and Sigri fields are awash with wild flowers in spring. These, with trees and bushes providing cover, can be excellent when passerine migration is underway. On one recent early May date the following were noted here: 750 European Bee-eaters, at least 20 Icterine Warblers and 110 Wood Warblers, Spotted (more than 250), Pied, Collared and Semi-collared Flycatchers, 15 Woodchat Shrikes and at least 185 Black-headed Buntings. Sigri is also good for hunting raptors. It has consistently turned up rarities, including the first Spectacled Warbler (an adult male) for Lesvos in mid-April 2011.

The Ipsilou area

Ipsilou Monastery is about 11 kilometres from Sigri, and the country between the two places is good for Eleonora's Falcon and Lesser Kestrel, which are thought to breed on offshore islands. During spring passage it is worth scrutinising the falcons for the Eurasian Hobbies and Red-footed Falcons that can sometimes join them. Ipsilou is one of the best migrant areas on the island in both spring and autumn and can witness good falls. Then, the trees on the steep slopes of the crag on which the monastery perches can be dripping with flycatcher and warblers. On 27 April 2008, for example, in a period with a north-westerly blowing, birders estimated more than 500 Blackcaps here, along with many other *Sylvia* and *Phylloscopus* warblers, Golden Orioles, four Collared, one Semi-collared and four Red-breasted Flycatchers, a Great

An adult Spur-winged Plover (*Vanellus spinosus*) at Faneromeni Fields in April 2011. This species, a scarce summer visitor to parts of south-east Europe, is now annual on Lesvos in the period March to May but is only very rarely seen in autumn.

Spotted Cuckoo and two Thrush Nightingales. The island's second Finsch's Wheatear (a male) was here in early April 2010 and the third in 2011. Also, the summit is a good vantage point from where to watch passing raptors and Pallid and Alpine Swifts. Be warned, though, it is a steep, and often hot, climb to the top from the road below

Meladia Valley

A road runs roughly south-east from Sigri towards Eresos through the dry, undulating countryside of the Meladia Valley. It does not look especially promising at first glance but appearances can be deceptive. The area around the ford, about halfway, has a proven track record for migrants, especially shrikes, warblers and flycatchers. It is worth checking any areas with cover or pools. In the first week of May 2009 more than 50 Red-backed and 10 Lesser Grey Shrikes were counted in the Meladia Valley. A flock of seven Blue-cheeked Bee-eaters – the only multi-bird record for the island – was

Rüppell's Warbler

Although BirdLife does not consider this *Sylvia* warbler to be threatened, its numbers are declining, and on Lesvos have been reduced to a handful of breeding pairs. Rüppell's Warbler breeds in scrubby habitats in Greece and large parts of western Turkey, with wintering grounds in coastal North Africa from Tunisia to Egypt and across large areas of Chad and Sudan.

Rüppell's Warbler breeds on Lesvos but it is a very localised summer visitor. The best-known location to see it is at Kavaki, between Petra and Molinos, where there is a handful of breeding pairs. Another former breeding site, at the south of the Mytilini peninsular, may no longer support a population. A few presumed migrants en route to Turkey are seen each spring, but records away from Kavaki have declined, reflecting the species' overall status. Kavaki is good for Eastern Orphean and Subalpine Warblers, and other migrants.

Thrush Nightingale (*Luscinia luscinia*) is a regular passage migrant, occurring in small numbers in April and May. It is virtually unknown in autumn.

seen here in early May 2008. And for additional spice, the breeding birds in this area include Chukar, Rufous Bush Robin, Western Rock Nuthatch, and Cretzschmar's and Cinereous Buntings.

Around Molivos

In the far north of the island, the area around the coastal town of Molivos is worthy of exploration. The town itself is a good place to look for passage raptors in spring and autumn, and just to the south is Perasma reservoir, an interesting site for water birds. Black-necked Grebe is possible in autumn, Ruddy Shelduck has bred and River Warbler has been noted on passage. The road running east from Molivos becomes a coastal track beyond Efthalou. Between there and Skala Sikaminias it passes small valleys with scattered trees on the north-facing slopes. These can be excellent for *Sylvia* warblers and other passerine migrants in spring and summer, and passage raptors and hirundines may pass overhead. Shearwaters may be on the move offshore, and the beautiful harbour at Skala Sikaminias sometimes attracts Audouin's Gulls, which breed in small numbers on offshore islands.

GEORGIA

Black Sea

Georgian plain

BATUMI WATCHPOINT

Batumi

Lesser Caucasus mtns

TURKEY

- **LOCATION**
  On the Black Sea coast of south-western Georgia

- **FLYWAY**
  Mediterranean–Black Sea

- **AUTUMN**
  Up to 35 species of raptors possible, including very large numbers of European Honey-buzzard, Steppe Buzzard, Black Kite, four species of harriers, Short-toed, Booted, Greater Spotted and Lesser Spotted Eagles; Levant and Common Sparrowhawks, Red-footed Falcon, Black and White Storks; European Bee-eater, European Roller.

- **KEY SITES**
  Sakhalvasho, Shuamta, Chorokhi Delta.

- **THREATS**
  Illegal shooting and trapping; ecotourism and education programme set up with hunters and children to diminish these effects.

Batumi, Georgia

While local people have always known about the phenomenal raptor migration over their villages each autumn, neither birders nor conservationists knew the scale of this movement until they began systematic counts in 2008.

It was the establishment of the Batumi Raptor Count (BRC) in that year that was finally able to quantify the movement. Since then, dedicated teams of counters have re-written the record books. Batumi is the Western Palearctic's single most important convergence zone for migratory raptors in autumn. Since 2008 it has witnessed the biggest single day-counts for European Honey-buzzard and Montagu's and Pallid Harriers, as well as the largest single-season counts for Black Kite and Montagu's, Western Marsh and Pallid Harriers. On 10 October 2012 it joined the ranks of the premier league of migration watch-sites with an autumn total breaking through the 1 million mark.

For 10 species – European Honey-buzzard, Black Kite, Pallid and Montagu's Harriers, Levant Sparrowhawk, Steppe Buzzard (sometimes considered to be the *vulpinus* subspecies of Common Buzzard), Lesser Spotted, Greater Spotted, Steppe and Booted Eagles – the numbers streaming through the area represent more than 1 per cent of their world populations. For European Honey-buzzard that figure could be more than 50 per cent. Some 35 species of birds of prey have been logged, some in huge numbers but some rare and unexpected gems.

The mountains of the Lesser Caucasus are a formidable barrier for migrating birds of prey.

138

Hundreds of thousands of European Honey-buzzards (*Pernis apivorus*) are observed passing the Batumi watchpoints during the last week of August and the first week of September each year.

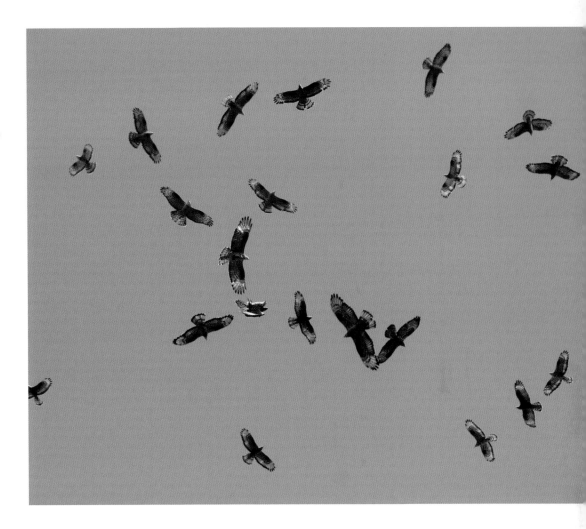

Reasons for the bottleneck

Many species move around the east of the Black Sea from breeding ranges in the western Caucasus, western Russia, Finland, Kazakhstan and probably the Baltic States. They include tens of millions of passerines and near-passerines, waders, storks and birds of prey. In the case of the last group it seems that many birds from north-east Europe also use this route to the Middle East and on to East Africa. The high massif of the Greater Caucasus presents a barrier, but there are passes through these mountains. To the south of the Georgian plain, the Lesser Caucasus rise more than 3,000 metres, with the summits often shrouded in cloud. These mountains, and probably especially the banks of cloud, present a huge obstacle, particularly for large, broad-winged birds such as buzzards, eagles and storks. Equally daunting is the vast expanse of the Black Sea, which dampens the formation of thermals. Consequently, unless weather conditions are excellent (that is, cloudless or with high cloud and only light winds), the easiest option for the birds is to squeeze around the western end of the Lesser Caucasus, south of the coastal town of Batumi, and thence into Turkey. Some continue to follow the Black Sea coast and others fan out to the south-east. Observations on the Turkish side of the mountains had pointed to this being a major bottleneck, but the BRC survey work led to much more accurate estimates of the numbers passing.

Record-breakers

Two count locations are organised, at Sakhalvasho and Shuamta, 2 kilometres and 6 kilometres from the coast, respectively. They are staffed by trained BRC volunteers, but visiting birders are also welcome. In 2012 they scored more than 1 million raptors, the best year to date. The season starts with a trickle in mid-August and quickly accelerates towards a first peak at the end of the month. Numbers remain high through September and early October. Among them, on 3 September 2012, were a record-breaking 179,342 European Honey-buzzards out of a season's total of 650,000 for this species. The last week of August and the first week of September is generally the busiest time for this species. This is also the period when the Montagu's Harriers peak.

From mid-month, while honey-buzzard numbers decline, their place is taken by ever-increasing tallies of Steppe Buzzards, which on 24 September 2012 stretched the counters to their limits: 56,527 (out of a season's total of almost 200,000) passed on that date. September also witnesses the movement of thousands of Levant Sparrowhawks (mostly early in the month), Western Marsh, Hen, Pallid and Montagu's Harriers, Black Kites, and Booted and Lesser Spotted Eagles, with Greater Spotted and Short-toed Eagles in good numbers. Of those, Black Kites are the most plentiful, with more than 100,000 being involved in 2012. That total represents the largest passage of the species ever observed in a single season. Raptor numbers fall in the second half of October.

Of all the raptors passing through Batumi in autumn, Steppe Buzzard (*Buteo buteo vulpinus*) is only outnumbered by European Honey-buzzard.

The unexpected

At the other end of the numerical scale in 2012 were 12 Oriental Honey-buzzards, 12 Eastern Imperial Eagles and singles of Cinereous Vulture and Saker. Small numbers of the first of these species have been detected regularly moving south at Batumi, begging the question – why? Some birds do migrate through Chokpak Pass in Kazakhstan, but that is far to the east and the nearest breeding population is in central Siberia. An Eleonora's Falcon is also an intriguing record.

Apart from the birds of prey, significant numbers of Black and White Storks and Common Cranes also use this route. Other species for which it is important in late August and September include hundreds of thousands of European Bee-eaters and many European Rollers; more than 1,800 of the latter were noted in 2012.

Not every day sees the floodgates open. If conditions are very good, with high pressure and an absence of cloud, more birds of prey probably to pass over the mountains. And if the weather is very wet and windy, they tend to remain grounded. Ideal conditions for a big movement are probably either just before the passage of a weather front bringing rain, or when conditions are improving after several days of poor weather.

Chorokhi Delta

Down on the coast, just a few kilometres away, the Chorokhi Delta has a mosaic of scrub, grassy plains, marshes and shoreline. These attract migrant herons, rails, waders, terns and passerines in the spring and autumn. In September 2009 two flocks of Sociable Lapwings were found here, and in 2012 there was a single White-tailed Lapwing. More likely waders are Dunlin, Sanderling, Temminck's and Little Stints, Terek, Broad-billed, Curlew and Wood Sandpipers, Common Greenshank and Ruff. Flocks of several hundred Black-winged Pratincoles typically stop over in late September. Good numbers of Calandra and Greater Short-toed Larks, Tawny, Richard's and Red-throated Pipits, Citrine Wagtail, Caspian Stonechat (the regional form of European Stonechat, *Saxicola maurus variegatus*, which may be a full species) and Eastern Black-eared and Desert Wheatears can all be seen in autumn.

Conservation

Since the numbers of raptors passing Batumi are so large (and hence at least reasonably representative), it is hoped that the data gained will indicate how the species' breeding populations are faring. This is particularly important since several of them are classified as Near Threatened (Pallid Harrier, Red-footed Falcon and Cinereous Vulture), Vulnerable (Eastern Imperial Eagle) and Endangered (Saker and Egyptian Vulture). Another important function of the programme is education and various projects are running to teach local people about the raptors and convince them they are a unique resource and worth conserving.

EGYPT | ISRAEL
JORDAN

KIBBUTZ ●
LOTAN

● **EILAT**

Sinai
Desert

Gulf of Aqaba

SAUDI
ARABIA

- **LOCATION**
  At the northern apex of the Gulf of Aqaba and the southern end of the Arava Valley

- **FLYWAY**
  East Asia–East Africa and Mediterranean–Black Sea

- **SPRING**
  Migration runs from February to early June and involves wildfowl, some seabirds, storks, raptors, waders, gulls, terns, swifts, bee-eaters, larks, pipits and wagtails, chats and thrushes, warblers, flycatchers and shrikes.

- **AUTUMN**
  Return migration starts with waders in late June and runs through to late November, with the last Common Cranes. The passage of large birds, notably storks and raptors, is much lighter in autumn.

- **KEY SITES**
  Eilat North Beach, parks, date plantations and Birdwatching Park; Ein-Netafin and Eilat Mountains; Arava Valley saltpans and pools; Yotvata; Kibbutz Lotan.

- **THREATS**
  Urban and agricultural development.

Eilat, Israel

Tens of millions of birds, including over a million birds of prey, fly over the southern Israeli resort of Eilat every spring. The species involved read like a 'who's who' of Western Palearctic migrants, and it is almost certainly the planet's best spring location for migrating raptors.

It is early April and first light. You decide to take a stroll around one of Eilat's parks before the mercury starts to climb and the crowds appear. A few contact calls alert you to the presence of birds. This is encouraging because the previous evening it was silent. A group of Squacco Herons, alarmed at your approach, fly out of a tree. On the grass are a few Tree Pipits, Yellow and White Wagtails and Northern Wheatears. Some buntings demand scrutiny: most are Ortolans but there are a couple of Cretzschmar's with them. To round off the pre-breakfast walk there are a couple of Whinchats and a Masked Shrike …

After breakfast, there is a change of habitat and a change of pace at some saltpans just north of the town. The noisy alarm calls of Spur-winged Lapwings punctuate your circuit around the pools, while the site's colour is provided by pink-hued Greater Flamingos. The pools hold a fine selection of waders: Black-winged Stilts, a couple of Collared Pratincoles, Kentish Plover, Dunlin, Little Stint, Wood, Green and Marsh Sandpipers, Common Redshank, Common Greenshank, Ruff and a single Broad-billed Sandpiper. A Red-necked Phalarope is performing its characteristic spinning routine. With the exception of the Spur-wingeds, none of these birds breed locally. They are on their way north, with destinations ranging from Turkey to the high Arctic. This, surely, is migrant-watching at its very best.

Avian crossroads

Eilat's geography dictates its excellence. The unimpressive resort town stands at the northern end of the Gulf of Aqaba, which runs to the Red Sea between Egypt to the west and Jordan and Saudi Arabia to the east. Either side of the gulf are mountains and desert, the Sinai Desert to the west and the Arabian Desert to the east. North of Eilat, the line of the Arava Valley, the Dead Sea and the Jordan Valley provide a low-lying corridor between more arid mountains. This line points towards Turkey, eastern Europe and Russia and is the route used by tens of millions of birds between February and early June and between late June and late November. In spring, most passage birds of prey fly north-east to the north of the town, crossing from the Sinai Desert over the Arava Valley and thence over the mountains of Jordan. More than a million take this route, though in autumn the vast majority follow a trajectory into the Sinai Desert further to the west.

While geography is the factor that determines the numbers and species passing over Eilat, weather (particularly adverse conditions) is important in determining how many of these birds are actually seen by human eyes, and how many stick around. For example, large birds of prey and storks that use thermals to help them cover great distances with the minimum expenditure of energy will be more reluctant to continue their journeys if poor weather suppresses the thermals. Fewer may be seen passing overhead, though more may be found grounded in the desert. Again, man-made modifications to the landscape are important in encouraging birds to linger, and the Eilat area has these aplenty: irrigated fields, salt pans, sewage works and freshwater pools. Wildfowl and waders take advantage of these to feed and rest. Meanwhile, passerines use the cover of town parks and date-palm plantations, and food available in kibbutz fields for the same reason. The combination of these factors makes Eilat one of the best migration hotspots on Earth.

Spring starts early

Spring migration lasts for several months. Steppe Eagles begin to pass as early as January, peaking towards March. They herald the beginning of a new spring, and by early March a wide range of migrants has already appeared. The peak period runs from mid-March to early May, though late northbound Masked Shrikes are still seen in early June. Early spring sees wintering birds (some finch species, for example) depart the area. As they leave, so the migration conveyor belt

Steppe Eagle (*Aquila nipalensis*) is one of the dominant spring raptors at Eilat, en route from Africa to Central Asia.

143

brings replacements from the south in March and April. The annual Eilat Birdwatching Festival is held at primetime, the end of March. During that period, it is possible to see 230 species in a week, with spectacular numbers to boot. There will be wildfowl, including Garganey; herons, including Black-crowned Night, Squacco and Purple Herons, Little and Great Bitterns; Glossy Ibis and Spoonbill; and White and Black Storks (day counts of more than 10,000 of the former are possible).

Raptor bonanza

However, raptors are Eilat's spring *piece de resistance*, passing through in astonishing numbers. More than 40 species have been recorded, although a few are exceedingly rare. The key players are Steppe Buzzard (the eastern, *vulpinus* subspecies of Common Buzzard) and – later in spring – European Honey-buzzard. Numbers of both approach or exceed half a million. Steppe Eagle and Black Kites number in the tens of thousands. Levant Sparrowhawks are late arrivals, streaming through in late April and early May (typically peaking around 25 April); flocks several hundred strong are regular. During the birdwatching festival, 25 species or more are seen.

Pools attract Spotted, Little and Baillon's Crakes, though the last are rare, along with many species of waders, of which up to 40 species can be expected in spring. These include such crowd-pleasers as Cream-coloured Courser, Collared and Black-winged Pratincoles, Caspian Plover (now annual, with three in 2011 and 13 during spring 2012) and the rare Terek Sandpiper.

The best of the rest

Skuas and Pallas's Gulls may be seen at Eilat's North Beach, along with other gulls and terns. Marsh terns also track up the Arava Valley, where they can be seen feeding over the pools. There may be thousands of White-winged Black Terns; 80,000 have been counted on a single day. European Turtle Dove, Common and Great Spotted Cuckoos, Scops Owl, European and Egyptian Nightjars and four species of swifts (Common, Pallid, Alpine and Little) all move through the area. Loose flocks of European Bee-eaters sometimes pass for hours on end, carrying with them their Blue-cheeked congeners (Little Green can also be seen, but they are residents).

Larks, martins and swallows, pipits and wagtails feature prominently. Greater Short-toed Lark, Red-throated Pipit and Yellow Wagtail are three of the commonest species of spring but for every common species there are the scarce: Bimaculated Lark, Olive-backed Pipit and Citrine Wagtail are regular. Any patch of roadside habitat, hotel lawn or acacia may hold chats (including Rufous-tailed Rock Thrush); one of at least 10 species of *Sylvia* warblers, not to mention the *Acrocephalus*, *Locustella* and *Phylloscopus* warblers. Five species of flycatchers are regular and then there are the shrikes, Golden Orioles, and Ortolan, Cretzschmar's, Cinereous and Black-headed Buntings.

By the end of April the bulk of the passerines have moved through but the

A male Bluethroat (*Luscinia svecica*). This chat is a winter visitor to Eilat and the Arava Valley, arriving in October and November and leaving for Russian and Central Asian breeding grounds in March and April. Long-term ringing studies have shown that most of the birds wintering in the area are males.

excitement is not over, since late-moving Levant Sparrowhawks, European Honey-buzzards, European Rollers and Masked Shrikes will still be moving. No two springs are the same. In March 2012, for example, there were exceptional numbers of Pied Wheatears, including two of the Asian subspecies *vittata*, the first for Israel.

Eilat's North Beach, at the head of the Gulf of Aqaba, is a strange combination of holiday resort and one of the planet's top migrant hotspots.

Summer and autumn

Eilat's North Beach can provide interest even in midsummer, in the short slot between the two migration seasons, but note that temperatures may hit 40°C or more in July, August and September. After southerlies, seabirds from the Indian Ocean sometimes appear, for example Cory's and Sooty Shearwaters. In late June 2011, Bridled and White-cheeked Terns and Long-tailed Skua were seen here, and the following June there was an Arctic Tern.

In autumn, far fewer raptors pass through but there are large numbers of many common species – and plenty of potential for finding rarities and scarcities. Passage wildfowl appear in the southern Arava Valley in early September and fields hold dozens of Greater Short-toed Larks and Yellow Wagtails and double-figure counts of Red-backed and Masked Shrikes. The first waves of European and Blue-cheeked Bee-eaters pass south. In the scrub, Willow and Eastern Olivaceous Warblers are plentiful. Later in the month the saltpans host growing numbers of Ruff, Little Stint, Dunlin, Marsh, Wood and Green Sandpipers and Caspian Terns. By the end of September the bee-eater flocks are seemingly never-ending, and there are large numbers of Northern and Isabelline Wheatears in any open areas. *Sylvia* and *Acrocephalus* warblers are now everywhere. In early October larks – including Desert, Bar-tailed

Desert, Temminck's and Thick-billed – are starting to congregate at favoured winter locations, and by the middle of the month large numbers of Bluethroats and Chiffchaffs are moving down the valley. The passage of European Bee-eaters and wagtails has all but dried up, but there are now more Tree and Red-throated Pipits. The season runs very late; for example, in November 2011 the Eilat area produced Yellow-browed and Menetries's Warblers, Olive-backed and Blyth's Pipits, Oriental Skylark, Steppe Grey Shrike, Rosy Starling, Little Bunting and Grey Hypocolius.

Eilat's parks, beach and saltpans

This modern resort is in the very hub of one of the world's greatest migration hotspots – so expect the unexpected. Even the lawns and trees of Ofira and Central parks can be alive with birds first thing in the morning before they are disturbed. Also in town, the Israeli Bird Ringing Centre (IBRC) has its own very productive ringing station, which has produced many rarities; these included, in the space of a fortnight in late August/early September 2012, Israel's eighth records of both Grasshopper Warbler and Paddyfield Warbler.

East of the marina, North Beach looks out over the Gulf of Aqaba, with

Spotted Redshank (*Tringa erythropus*) and many other passage waders rest and feed at pools at Eilat and Yotvata.

its gulls, terns and maybe something rarer. Scan the sea and shore for migrant wildfowl, herons, gulls and terns among the resident Western Reef and Striated Herons, White-eyed Gulls and Pied Kingfishers. Cory's Shearwaters are sometimes seen in late summer. And 'megas' seen from here have included the Western Palearctic's first Mascarene Shearwater (in 1992) and Lesser Frigatebird (in 1997), Streaked Shearwater and Red-billed Tropicbird.

Towards the Jordanian border, a sewage canal runs inland from North Beach; the south (beach) end of the canal attracts waders, including Greater Sand Plover, and herons and crakes. The reedier areas are better for crakes, but only if there is water in the canal. Inland, west of the canal, the salt pans are good for waders, marsh terns, Citrine Wagtail and Red-throated Pipit. On the north side of town, the cemetery is worth checking; Israel's second Red-flanked Bluetail was there in November 2011, and the bushes attract

Ringing recoveries of Levant Sparrowhawks (*Accipiter brevipes*) passing through Eilat end up as far away as Romania, Ukraine, Russia and Syria. The peak migration period is the third week of April when many arrive at roost sites well after nightfall.

Eastern Olivaceous Warblers (*Hippolais pallida*) are plentiful in April and May. The first returning birds arrive in early August, and they are again numerous in September.

commoner chats and warblers. This is also a good place from where to watch raptor migration in the autumn; although the numbers involved are nowhere near those of spring, they can still be impressive. North again, and running alongside Highway 90, are the Eilot Fields, which in recent times have attracted Black-winged Kite (in late August 2011), the Siberian subspecies of Buff-bellied Pipit (late November of the same year) and two Caspian Plover in March 2012.

Eilat Mountains

Overlooking Eilat from the west are several raptor watchpoints close to the road to Ovda. The first (the 'low mountain') is about 2 kilometres west of the town centre. Then there are several off-road viewpoints near the 734-metre Mount Yoash, including Ein-Netafin. When they are on the move, Black and White Storks, European Honey-buzzard, Black Kite, Short-toed Eagle, European and Levant Sparrowhawks, Steppe Buzzard, Lesser Spotted, Steppe and Eastern Imperial Eagles and Common Crane can all be seen from here. With a big slice of luck, you could even luck out with an Oriental Honey-buzzard, though the chances of a repeat Bateleur or Lappet-faced Vulture fly-over are slim indeed.

The best periods are around mid-morning and late afternoon, since that is when the large raptors are able to get sufficient lift. Numbers are smaller first thing and when the sun is going down when the thermals are not developing. In the hottest part of the day they will be too high to see.

Male Black-headed Bunting (*Emberiza melanocephala*). This species breeds in northern Israel and winters in India, so the main migration route is north of Eilat. However, small numbers can be seen in May and September.

The mountain wadis here are good in spring for Rüppell's and Eastern Orphean Warblers.

North to Yotvata

The Arava Valley runs between the Edom Mountains in Jordan to the east and the southern Negev Desert to the west. Highway 90 follows the valley, passing a succession of sites whose names are the stuff of legend. There are the saltpans and freshwater pools of Km19 (19 kilometres north of Eilat) and Km20, which are guaranteed to pull in waders and marsh terns. The Western Palearctic's first Southern Pochard was at Km20 in April and May 1998, and more recently this area has had Common Goldeneye, White-tailed Lapwing, Caspian Plover, Terek Sandpiper, Rufous Turtle Dove and Menetries's Warbler.

Around 40 kilometres north of the resort, the agricultural fields, acacia scrub, date-palm plantations and sewage works of the Yotvata area pull an astonishing number of migrants. In a typical spring these will include Purple Heron, Eurasian Teal, Mallard, Garganey, Egyptian Vulture, Steppe Eagle, Eastern Imperial Eagle, Osprey, Lesser Kestrel, Ruff, Common Snipe, Green Sandpiper, Black-headed Gull, Common and Alpine Swifts, Eurasian Wryneck, Oriental Skylark, House, Sand and Eurasian Crag Martins, Barn and Red-rumped Swallows, Richard's, Tawny and Red-throated Pipits, Pied, Black-

Menetries's Warbler (*Sylvia mystacea*) is a rare passage migrant in the Arava Valley, most often recorded in March and November. This bird is a male.

headed (*Motacilla flava feldegg*) and Citrine Wagtails, Common Nightingale, Bluethroat, many warblers, Woodchat and Masked Shrikes and several species of buntings. In November 2010 two Black-crowned Sparrow Larks were here for a fornight. In 2012 Yotvata scored with a spring Sociable Lapwing and Caspian Plover, and autumn White-tailed Lapwing. Egyptian Nightjars can be seen hawking over Yotvata fields at dusk. This is not a site to be missed!

Then there is the world-famous Km33, which has justifiably been described as the "best site in the Western Palearctic for larks and wheatears" and has had Dunn's Lark and Red-tailed Wheatear among numerous exciting resident and migrant species. Bar-tailed Desert, Desert, Hoopoe, Lesser Short-toed,

Red-rumped Swallow (*Cecropis daurica*) is a common passage migrant in spring and autumn and can be seen anywhere in the valley.

Steppe Eagle

Steppe Eagle (*Aquila nipalensis*) is the largest of the raptors to come through Eilat in big numbers. This magnificent bird has a wingspan up to 1.9 metres. Although BirdLife does not consider it to be endangered, there is no doubt that it has experienced a decline in population, something reflected in the numbers passing through Israel each year. No one has an accurate idea of its global population, with estimates put at "more than 10,000 individuals". The breeding range is certainly large, spanning much of Central Asia and China, but it is thinly spread and the former breeding population in Turkey has been virtually extirpated. Birds that pass through southern Israel are from the East African wintering population. Others spend the winter in Iraq, South Asia or South-east Asia. Threats to the species include the conversion of steppe to agricultural land; persecution; and collisions with power lines.

Bimaculated, Temminck's and Thick-billed Larks have all been seen here as have Desert, Mourning and Hooded Wheatears, Tristram's Starling and Desert Finch. Between November and March this can also be a good site for Asian Desert Warbler.

Kibbutz Lotan

About 50 kilometres north of Eilat, Kibbutz Lotan and the adjacent Lotan Bird Reserve have established a fine reputation in recent years. The kibbutz is managed with a solid ecological ethos to be as attractive as possible to migrant birds. There are gardens with lawns, shrubs and trees, as well as agricultural fields. This desert oasis has a bird-list of 275 species. Regular, generally common, migrants include Common Quail, Eurasian Wryneck, Tawny and Red-throated Pipits, Rufous Bush Robin, Common Nightingale, Black-eared and Isabelline Wheatears, Rufous-tailed Rock Thrush, Rüppell's, Eastern Bonelli's, Eastern Olivaceous, Eastern Orphean, Barred and Olive-tree Warblers, Collared and Semi-collared Flycatchers, Golden Oriole, Red-backed Shrike, Ortolan and Cretzschmar's Buntings. And, of course, there are the overflying storks and raptors. Kibbutz members established the Lotan Bird Reserve in 1996 on what had previously been sandy, barren desert. It has a ringing station; the reserve's reedy pool is a magnet for passage Great Reed and Savi's Warblers; wheatears and larks visit a large drinking pool; raptors sometimes punctuate their migration in taller trees; and the alfalfa field lures Richard's Pipits in October and November. Israel's second Dusky Warbler was here in 1996 as was the country's fourth Pied Stonechat in November the following year.

EGYPT

Sinai Desert

Gulf of Suez

Gulf of Aqaba

JORDAN

SHARM
EL-SHEIKH

RAS
MUHAMMED

Red Sea

- **LOCATION**

  *Southern tip of Sinai
  Desert*

- **FLYWAY**

  *East Asia–East Africa and
  Mediterranean–Black Sea*

- **SPRING**

  *Waders, raptors and
  passerines.*

- **AUTUMN**

  *White Storks provide
  the most spectacular
  migration, along with
  large numbers of raptors;
  also, waders, bee-eaters,
  hirundines, pipits and
  wagtails, warblers,
  flycatchers and shrikes.*

- **KEY SITES**

  *Ras Muhammed National
  Park, El Qa Plain, Sharm-
  el-Sheikh sewage works,
  Nabq Protected Area.*

- **THREATS**

  *Tourist development;
  leisure pursuits offshore.*

Sharm el-Sheikh, Egypt

*As the mid-morning temperature begins to soar in early
autumn, so do the birds. Hundreds, then thousands of White
Storks lift off from the desert around Sharm and gain height
for their big sea crossing to the African continent.*

The Egyptian resort of Sharm-el-Sheikh is probably best known as a great holiday destination, especially for its coral reefs and snorkelling. It is also close to where geography dictates that most of the world's White Storks make their autumn crossing of the Red Sea. Sharm is close to the southern apex of Sinai, the triangular block of desert between the Gulf of Suez and the Gulf of Aqaba, two relatively narrow stretches of sea that converge to form the Red Sea. Their arrangement does much to dictate the movement of large soaring birds in the region because, together with Suez to the north and the Straits of Bab al-Mandeb (between Djibouti and Yemen) to the south, this is one of the great crossing places for birds between Africa and Asia.

Different strategies

Especially in autumn, millions use the line of the Arava Valley and Gulf of Aqaba, or track straight across the desert, on their way south. For the storks and large raptors the Red Sea is a formidable obstacle. Many of the birds of prey pass north of Suez to avoid having to cross the Red Sea and then either follow the Nile Valley or the west coast of the Red Sea. Satellite telemetry has shed light on the movement of raptors, including the lengths that some go to in order to avoid sea crossings. Two autumns running a Lesser Spotted Eagle flew to the southern end of Sinai but then moved back north to Suez rather than make the Red Sea crossing at Sharm; this long detour involved three days and 500 kilometres of extra flying time. Others, following a more eastern route out of Asia, track down the east of the Red Sea and cross into Africa over the Straits of Bab al-Mandeb. However, a minority of large raptors and huge numbers of White Storks do use the Sharm route.

In contrast with the arid, inhospitable Sinai Desert to the north, Sharm el-Sheikh has hotel gardens with shrubs and trees; and its golf courses have well-watered greens and fairways. The area also has freshwater pools, a sewage works and an intertidal zone. In short, it is perfectly suited to cater for large numbers of stop-over migrants.

Up to half a million White Storks (*Ciconia ciconia*) pass through Sharm in August and September before crossing the Red Sea to the coast around Hurghada. In September they are accompanied by tens of thousands of raptors.

Spring migration

By early April migration is in full flow, although the number of birds involved is not as great as in autumn. A good mix of waders can be found feeding along the shoreline at Ras Muhammed National Park, at Sharm sewage works and at the golf course pools. Passage waders – Collared Pratincole, Common Ringed Plover, Little Stint, Dunlin, Ruff, Whimbrel, Eurasian Curlew, Common, Green and Wood Sandpipers, Common Greenshank, Common Redshank and Ruddy Turnstone – join noisy breeding Black-winged Stilt and Spur-winged Plover to feed before moving on.

Any lawn, hotel garden shrubbery or patch of cover in the desert may hold Common Quail, European Turtle Dove, Eurasian Wryneck, pipits, wagtails, chats, warblers and buntings. Fewer large soaring birds pass over Sharm in spring, but good numbers are still encountered. At the start of April 2005, for example, one observer witnessed a flock of 1,000 Steppe Buzzards passing over Ras Muhammed. Lesser Kestrels move north between mid-March and mid-April, and Sooty Falcons arrive from their Madagascan wintering grounds in late April. Rarities are possible; in March 2009 Egypt's second-ever Grey Hypocolius was at Nabq, just north of Sharm.

'River of Storks'

The phenomenal movement of White Storks begins in August though the flow does not finish until late October. Survey work in 1998 logged more than 275,000 of the species passing the Sharm area in August and September, though it was estimated the true figure was probably 390,000–470,000. Many rest on sandy beaches, especially on the Gulf of Suez side and also in the surrounding desert, before attempting the crossing. In the height of the season an average of around 12,000 a day arrive, rest and move on. Most fly a little further north, to the coastal mountain ridge of El Tor, from where they cross the Gulf. Although raptor numbers at Sharm in autumn do not compare with some other bottlenecks, there is still a broad range. Black Kites peak in early September, followed by Levant Sparrowhawks (albeit in much smaller numbers), Steppe Buzzards, harriers, Ospreys and Lesser Kestrels. Great White Pelican, Black Stork, and Short-toed, Lesser Spotted and Eastern Imperial Eagles are also seen.

Black-necked Grebe, Garganey, Eurasian Teal, Northern Pintail, Tufted Duck and waders arrive at the sewage works and other freshwater pools; Greater Sand Plover and Grey Plover return to winter in the intertidal zone; European and Blue-cheeked Bee-eaters pass over in large numbers, as do hirundines; and passerine passage extends from late August to early November. Eurasian Wryneck, Greater Short-toed Lark, Tree, Tawny and Red-throated Pipits, Yellow and White Wagtails, Northern, Isabelline and Eastern Black-eared Wheatears, Common Redstart, Common Nightingale, Whinchat, Bluethroats, Savi's and Eastern Olivaceous Warblers, along with a range of *Acrocephalus*, *Sylvia* and *Phylloscopus* warblers, and Red-backed, Woodchat and Masked Shrikes may be seen. The qualifier 'in suitable habitat' does not really apply. For example, a Savi's Warbler has been seen hopping between plant pots in a plant nursery in Sharm el-Sheikh. Rarities have occurred regularly in recent autumns, including the Western Palearctic's first Streak-throated Swallow (November 2003), Egypt's first Pin-tailed Sandgrouse since 1919 (December 2004); and a Pintail or Swinhoe's Snipe (October 2008).

A small number of Long-legged Buzzards (*Buteo rufinus*) winter around Sharm el-Sheikh. The species breeds not far away, in Jordan and Israel.

Where the desert meets the ocean. Cliffs of fossilised coral loom over the Red Sea, with arid, sun-baked Ras Muhammed National Park beyond. This area appears inhospitable, but the shoreline provides feeding for herons and waders; even the smallest shaded areas shelter passerines; and storks and raptors rest here before making the crossing to Africa.

Exploring the area

At the very southern tip of the Sinai Desert, 15 kilometres south-west of Sharm, Ras Muhammed National Park comprises ancient coral reefs that have been raised above the level of the Red Sea, sandy bays, intertidal flats and a small area of mangroves. Away from the coast are gravel and sand plains, morphing into sandstone mountains and dissected by wadis. The area is a hot, sparsely vegetated and uncompromising landscape. The intertidal zone is good for herons and waders; gulls and terns breed or winter offshore; and any dwarf acacia or shadow behind a rock can provide welcome shade for migrants.

A few kilometres north of Ras Muhammed, near El Tor, is the southern expanse of El Qa Plain, Here, large numbers of storks and raptors can sometimes be seen resting on the desert. They then use a steep-sided coastal ridge to gain altitude for the crossing over the Gulf of Suez. When systematic counts were conducted in autumn 1998 it was found that 70 per cent of the White Storks counted at Ras Muhammed actually crossed the sea near El Tor.

Sharm el-Sheikh itself has much good habitat. The sewage pools and their adjacent plantation and scrub are excellent in spring and autumn. In addition to migrants they also attract sandgrouse, including Lichtenstein's, to drink in the evening. Interesting raptors can be seen here very late in the autumn, including wintering Long-legged Buzzard. Two Cinereous Vultures (a rare passage migrant) were here in late November 2008. The new pools west of Na'ama Bay and those at the golf resort are good for waders, while the fairways and greens attract larks, pipits, wagtails and wheatears. The combination of watered lawns and shaded gardens also make hotel grounds good places to search for migrants. And the intertidal zone at Nabq Protected Area, north of the town, has waders. Nabq also has probably the most northerly mangroves in the world.

LAKE BARINGO

LAKE BOGORIA

LAKE NAKURU

Rift Valley

LAKE NAIVASHA

KENYA

Nairobi

- **LOCATION**
  Central Kenya

- **FLYWAY**
  East Asian–East African and Black Sea–Mediterranean

- **SPRING**
  Palearctic-bound migrants pass from February to May, including wildfowl, raptors, waders, terns, hirundines and other passerines; intra-African migrants arrive from southern tropical Africa from February to May.

- **AUTUMN**
  Between late August and November, Palearctic breeders return for the winter; northern tropical African migrants arrive or pass through between July and September, including Grasshopper Buzzard and White-throated Bee-eater.

- **KEY SITES**
  Lakes Baringo, Bogoria, Nakuru and Naivasha.

- **THREATS**
  Lakeside development, agricultural pollution, invasive species.

Rift Valley, Kenya

The soda lakes of the East African Rift Valley conjure up images of pinks clouds of foraging flamingos, one of the greatest wonders of the natural world. Less well known but equally important, the valley's wetlands, savanna and forest supports millions of Palearctic and tropical African migrants.

The Rift Valley runs roughly north-south from the Red Sea coast through the mountains of Ethiopia before dividing. The western, Albertine Rift runs south through Uganda then separates Tanzania from the Democratic Republic of Congo. The Eastern Rift slices like a giant knife through the Kenyan mountains as it crosses the equator. A series of shallow lakes – Baringo, Bogoria, Nakuru and Naivasha – lie on the floor of the valley, two great mountain blocks rising to the east and west.

The importance of the area for migrant birds cannot be underestimated. Of Kenya's 1,100 bird species, about 170 are Palearctic-breeding migrants, many of which spend part of their year in the Rift Valley. Some just pass through, while other linger for several months. Several dozen other species are intra-African migrants.

The valley's mixture of freshwater and alkaline lakes, muddy margins, savanna and acacia and other woodland supports millions of birds throughout the year. These range from Nakuru's famed Greater and Lesser Flamingos to a bewildering array of Afro-tropical species: hornbills, wood-hoopoes, bee-eaters, larks, bush-shrikes, cuckoo-shrikes, cisticolas, starlings, weavers and many others.

The abundance of food, particularly invertebrate prey, ensures that the region also receives millions of avian invaders. However, unlike some migration hotspots, there are no simple patterns here. Palearctic migrants arrive from August to November, their fat levels depleted after long, sometimes very long, journeys from the north. Some are in transit to destinations even further south, while others – notably large numbers of wildfowl and waders – take advantage of the feeding opportunities to spend the winter in the valley.

They are joined by intra-African migrants that may have come hundreds, rather than thousands, of kilometres from the north. These and the Palearctic breeders leave in spring, to be replaced by more birds that have bred south of the equator in the Austral summer and are seeking warmer conditions. As if this was not complex enough, some birds move according to where the rains are. Other species have both resident and migratory populations. Even the

The Rift Valley soda lake of Nakuru has supported up to 1.5 million Lesser Flamingos (*Phoenicopterus minor*), which form a pink band around its shores. It is also a prime migrant stopover.

flamingos of the soda lakes, while not true migrants, move from lake to lake. Consequently, and bewilderingly, there is an almost constant coming and going of bird species throughout the year.

The difficulty of defining seasons in this most equatorial of regions is easy to understand when one considers the distances that some birds have travelled to get here. Many of the Palearctic migrants are waders that have bred in the high Arctic. Others have come from the vast taiga forests of Eurasia, the mixed woodlands of Europe or the Mediterranean maquis. Some Barn Swallows, Northern Wheatears and *yakutensis* Willow Warblers will have travelled from the far north-east of Siberia.

Some migrants make repeated stops en route south in autumn and may take five months to reach their final 'destination' – not long before starting the return journey!

Autumn arrivals and transients

Large numbers of Palearctic wildfowl appear on the Rift Valley lakes in early autumn. Some move on after a short stay, while others remain until spring. Large numbers of Garganey, Eurasian Wigeon, Eurasian Teal, Northern Pintail,

Northern Shoveler and Tufted Duck are among the most plentiful of them. Ferruginous Duck, Common Pochard and Gadwall are sometimes seen but they are rare. Eurasian Spoonbill and Black and White Storks also pass south.

Raptor migration does not really get going until October when Pallid, Montagu's and Western Marsh Harriers appear, remaining until early April. Other passage and wintering raptors include Steppe Buzzard, Lesser Spotted, Steppe and Booted Eagle, Eurasian Hobby, Sooty Falcon, Peregrine and Lesser and Common Kestrels. Large numbers of Amur Falcons move through Tsavo, further east, in late October and early November. European Honey-buzzard, Greater Spotted Eagle, Saker and Eleonora's Falcon occur only rarely. A Levant Sparrowhawk was seen at Lake Baringo in November 2010.

Grasshopper Buzzard (*Butastur rufipennis*) is a small, mostly insectivorous predator. It is a migrant, moving south from Sudan and Ethiopia, where it breeds, into the Rift Valley in September.

Common Greenshank (*Tringa nebularia*) is a long-distance migrant breeding as far north as the Arctic and wintering south to South Africa. Some spend the winter in the Rift Valley; others use it on passage only.

Some Palearctic waders, Marsh and Green Sandpipers, for example, arrive in August. During September and October others include Caspian, Common Ringed, Little Ringed and Kentish Plovers, Curlew Sandpiper, Ruff, Jack and Common Snipe, Black-tailed Godwit, Spotted Redshank, Common Greenshank and Wood and Common Sandpipers. Great Snipe head further south, passing through the valley between October and December (and returning northbound in April and May). Rarer waders to look for are Long-toed Stint, Sanderling, Broad-billed Sandpiper, Pintail Snipe, Bar-tailed Godwit, Eurasian Curlew, Terek Sandpiper and Red-necked Phalarope.

European Nightjars pass in October and November and late March and early April. In late August the movement of Barn Swallows begins, followed in September by Common Swift, European and Blue-cheeked Bee-eaters, Sand and House Martins, Tree and Red-throated Pipits, Yellow Wagtail, Whinchat, Northern Wheatear, Willow Warbler and Red-backed Shrike. Great Spotted Cuckoo, European Roller, the *epops* subspecies of Hoopoe, Isabelline Wheatear, Rufous-tailed Rock Thrush, Spotted Flycatcher, Eastern Olivaceous and Barred Warblers follow in October, with Rufous Bush Robin one of the latest arrivals. By November there will be very large numbers of European Roller, Barred Warbler and Common Whitethroat in the valley.

Spring

White Storks use Lake Nakuru as a staging post on their way north in early spring. There were 5,000 there in late February 2010, for example. Rare migrant raptors may be seen in spring as well, in recent years including Long-legged Buzzard in January and February, Amur Falcon in February and European Honey-buzzard in April. Return wader passage increases the chance of scarcer species turning up among the commoner wintering birds. In recent years these have included Red-necked Phalarope and, in 2009, a Lesser Sand Plover at Lake Oloidien and a Pectoral Sandpiper in the same lake complex, at Lake Naivasha. In early April flocks of northbound European Roller and European Bee-eater can be seen passing overhead, and Red-backed Shrikes will be on the move.

Intra-African migrants

The pattern of intra-African migration is complex and far from well understood, but there are a number of different models. Grasshopper Buzzard exhibits one. It breeds in the Sahel and other arid grasslands south of the Sahara Desert in a band running from northern Ethiopia to The Gambia between March and May. After the breeding season the birds migrate south, some to the Kenyan Rift Valley, where they arrive in September to take advantage of the rich post-rains populations of invertebrates (particularly grasshoppers) and small reptiles. They return north early the following year. Wahlberg's Eagles also move into the Rift Valley and eastern Kenya in August, remaining until April when they return to breeding areas to the north, west and south of the country. White-throated Bee-eater has a similar migratory pattern, arriving in September from semi-desert areas to the north and departing in April.

The spectacular Pennant-winged Nightjar has a different migration strategy. It breeds in southern tropical Africa and is a double passage migrant in the Rift Valley, moving north in January and February to subtropical savanna from Sudan to Nigeria, then returning south again to its breeding grounds from July to early September.

Different again are Black and Levaillant's Cuckoos, which are present – and breed – locally between September and April but spend the winter in the forests of the Congo Basin. Jacobin Cuckoo has a resident population, but some birds of the southern African subspecies *serratus* are non-breeding visitors to the valley. African Paradise-Flycatchers breed in the southern

Willow Warbler

Willow Warbler (*Phylloscopus trochilus*) is a very familiar species with a number of claims to fame. Perhaps unsurprisingly it is the commonest summer visitor to Eurasia, with a population of between 300 million and 1.2 billion. It is thought to make up more than 15 per cent of the near-passerines and passerines moving between Eurasia and Africa. Less well known is the fact that one population of this 10g waif makes the longest migration of any songbird. Birds of the *yakutensis* subspecies breed in the far north-east of Siberia but spend the winter in southern Africa. They cross much of Siberia, Kazakhstan, Iran and Saudi Arabia before crossing the Red Sea and entering Ethiopia, from whence they move south through the Rift. Remarkably, some end up in southern Africa after a trip of 11,300km. Despite Willow Warbler's huge range and population, there have been worrying long-term declines in parts of its range, including Britain.

Willow Warblers (*Phylloscopus trochilus*) of the Siberian subspecies *yakutensis* undertake one of the longest migrantions of any passerine. On passage they are sometimes seen in very unlikely locations.

tropics during the austral summer (between October and March), then move north to Kenya. In addition to the true intra-African migrants there are many nomadic species, whose movements are related to rainfall and the supply of food and consequently are highly unpredictable. Examples include species as varied as Lesser Moorhen, Magpie Starling and Red-billed Quelea. Capped Wheatear is an altitudinal migrant, with non-breeding birds descending to the valley between April and September.

Freshwater lakes

The two northernmost lakes, Baringo and Bogoria, are both shallow water bodies between 970–990 metres above sea level but in other respects they are very different. Baringo is about 110km north of Nakuru town. It is a freshwater lake and at 16,800 ha, vast. Baringo is fed by several rivers and is just 6m deep. Soil erosion means it is getting shallower but it can become choppy in windy weather. The Tugen Hills rise to the west and the Laikipia escarpment to the east. Surrounding the lake is superb habitat for birds: acacia woodland, thicker bush, grassland and, especially where the rivers enter it, stands of reed and marsh grass.

More than 470 species of birds have been recorded here, and it is possible to see 200 of these in a single day. Lesser Flamingo is an occasional visitor, usually on passage; Madagascar Pond-heron is a rare non-breeding visitor; Pallid Harrier is a regular passage migrant, as are small flocks of Lesser Kestrel. To the east of the lake, the Solion Plains are good for grassland birds, including Caspian Plover in winter.

White-throated Bee-eater (*Merops albicollis*) arrives from the north in September. This species breeds in a narrow belt of arid country south of the Sahara Desert. Like several other intra-African migrants that winter in Kenya, it flies south after the breeding season and takes up residence in acacia woodland (or, further west, in rainforest). The birds move north again in April.

Ramsar-listed Lake Naivasha is about 80km north-west of Nairobi and easily reached from the city. Like Lake Baringo, this is a shallow freshwater lake, but it is much higher, about 1,900 metres. It is fed by the Malewa and Gilgil rivers but unusually has no outlet. Nearby is a big rose-growing area and there have been problems both with chemical run-off from the floriculture, and also with invasive plant species, notably water hyacinths. Another problem has been water abstraction for this industry, which has lowered water levels. International conservation organisations and local naturalists are working to get better protection for this magnificent site. The main lake covers 15,000 hectares and is about 8 metres deep, incorporating a partially submerged volcanic crater, which contains the deeper Crescent Island Lagoon. To the south-east and separated from the main lake by papyrus swamp and acacia woodland, lies Lake Oloidien, which is more alkaline.

More than 350 species have been recorded here, and surveys regularly note more than 80 kinds of waterbirds. Many Palearctic wildfowl visit, including Eurasian Wigeon, Eurasian Teal, Garganey, Northern Pintail, Northern Shoveler and Tufted Duck. Wildfowl numbers reach their peak in November and February, presumably as wintering birds are joined by passage migrants. Its margins are also good for wintering and passage waders and wintering Sedge and Basra Reed Warblers. There is a May record of Pectoral Sandpiper, a New World vagrant. The surrounding bush holds many migrant and resident passerines and near-passerines, including intra-African migrant cuckoos.

The soda lakes

A little south of Baringo, Lake Bogoria is a shallow alkaline body of water of 4,000ha, which supports phenomenal numbers of Lesser Flamingos and other water birds. On the west shore are geysers and hot springs, while the Siracho escarpment rises sharply from the east of the lake. Acacia woodland surrounds much of the lake. At the northern end is the Kisibor Swamp, which is a large freshwater wetland.

Lesser Flamingos are the dominant water birds, their numbers estimated to have reached 2 million in the recent past. The margins supports passage waders, the surrounding woodland has Palearctic migrants during the winter months, and both Pallid Harrier and Lesser Kestrel are regular passage migrants.

Lake Nakuru has featured in countless television documentaries, primarily because of its flamingos (there are small numbers of Greater Flamingos). Three major rivers flow into this very shallow, very alkaline lake, which covers 3,000ha and is bounded by hills, acacia woodland and – to the north – the town of the same name. Since fish were introduced in the early 1960s water birds other than flamingos have thrived.

About 450 species have been recorded here. There are large numbers of wildfowl, waders and other water birds. In winter there are good numbers of Little and Black-necked Grebes, Lesser and Greater Flamingos, African Spoonbill, Madagascar Pond-heron, Great White Pelican, Black-winged Stilt, Grey-headed Gull and Gull-billed Tern. Expect to find Ringed and Kentish Plovers, Ruff, Common Snipe, Black-tailed Godwit, Spotted Redshank, Marsh Sandpiper, Common Greenshank, and Green, Wood and Common Sandpipers here.

Magpie Starling
(*Speculipastor bicolor*)
**is a nomadic wanderer,
rather than a true migrant.
An inhabitant of thorn
scrub and dry bush in the
northern part of the Rift
Valley, its range fluctuates
from year to year.**

Lake Balqash

CHOKPAK PASS

Taukum Desert

Sorbulak Lake

● Almaty

Zhotasy mtns

Kyrgyz mtns

KYRGYZSTAN

- **LOCATION**

  Foothills of Tien Shan Mountains and steppes of central Kazakhstan

- **FLYWAY**

  Central Asian

- **SPRING**

  Prolonged, starting with wildfowl from late February and also involving herons, raptors, gulls, bee-eaters, larks and other passerines. The last migrants are seen in early June.

- **AUTUMN**

  Waders begin to appear from the north in July, their numbers growing in August (especially huge numbers of Red-necked Phalaropes). All the bird groups involved in spring reappear in the period until late October.

- **KEY SITES**

  Chokpak Pass, Korgalzhyn State Nature Reserve, Taukum Desert, Sorbulak Lake.

- **CONSERVATION**

  Modern agriculture, including cultivation of the steppes, overgrazing, mineral extraction.

Chokpak and Korgalzhyn, Kazakhstan

Kazakhstan boasts some of the biggest concentrations of waders on Earth in spring – and some spectacular visible migration through a mountain pass in autumn.

Tens of millions of birds pass through Kazakhstan on a broad front in spring and autumn on the Central Asian flyway. This route is taken by birds migrating between their wintering ranges in Africa, the Middle East and South Asia and their breeding grounds in Central Asia, Russia and Siberia. The value of migration watchpoints in Kazakhstan differs between spring and autumn. In spring, after a comparatively wet winter period, the steppes and semi-deserts are much more productive than they are later in the year. Seasonal lakes hold water – and billions of insects – and grasslands are similarly productive, offering good foraging possibilities for waders, passerines and some birds of prey.

The spring passage of Red-necked Phalaropes through the wetlands of central Kazakhstan is fairly typical. They begin to arrive in southern

Collared Pratincole (*Glareola pratincola*) is a common breeding migrant, nesting in wetlands in semi-desert areas. It arrives in mid-April or early May and leaves from mid-June to September.

Red-necked Phalaropes (*Phalaropus lobatus*) feeding on invertebrates on a Kazakh lake. The birds use the lakes as stopover sites in spring (between April and mid-June) and autumn. Adults predominate in July and August, with juveniles becoming the majority in late August and September. The numbers involved can reach hundreds of thousands in the Korgalzhyn complex.

areas such as Chardara Lake (south-west of Chokpak Pass) in April, and later in the month they can be seen, for example, at the Tengiz-Korgalzhyn wetland complex of central Kazakhstan. Sometimes they travel in small groups, but on occasion they can be seen in spectacular flocks of more than 200 birds. The most intense migration is in May when flocks of thousands may be observed. Between 589,000 and 653,000 have passed through Tengiz-Korgalzhyn in a single spring. Passage then tails off and the latest spring birds are recorded in mid-June. Autumn migration starts not long after – in July, when adults (especially females) dominate. From August onwards, juvenile phalaropes become more common than adults.

In autumn many southbound migrants make a detour around the north and west of the Caspian Sea to avoid the now-arid zones of much of Kazakhstan, the largest area of dry steppe on the planet. It is simply too dry to offer refuelling possibilities for energy-hungry birds. Using Red-necked Phalarope as an example, again, the peak autumn counts at Tengiz-Korgalzhyn run to hundreds of thousands. Most leave in September but the latest do not depart until mid-October.

Kazakhstan through the seasons

Wildfowl, Greater Flamingos and some species of larks are some of the first birds to start moving in late February. The first waders, more wildfowl,

Pin-tailed Sandgrouse and ever-growing numbers of passerines follow from early March. However, the period from early April to mid-May is the most concentrated, with wildfowl, herons, raptors, waders, gulls and terns, larks, chats, warblers, shrikes, finches and much else on the move. Stragglers, maybe including northbound White-winged Black Terns and European Honey-buzzards, may be seen on passage in early June. The wetland sites are best for wildfowl, herons and waders. Raptors, bee-eaters and ground-feeding birds can be seen pretty much anywhere.

Autumn migration is already underway in the second half of July. Those lakes that have retained their water and muddy margins will attract stopover waders, so the wetlands of Tengiz-Korgalzhyn are still good. Migration gathers force in August but September is the most exciting month. Chokpak Pass is the place to be for the visible migration of anything from raptors to pigeons, bee-eaters to hirundines, corvids and pretty much any other migratory passerine species. Meanwhile, wildfowl arrive from the north, many of them ending up at Lake Chardara, south-east of the mountains, where they winter. Autumn migration continues well into November.

Dry valleys on the northern slopes of the Tien Shan foothills offer additional cover and can become migrant traps.

Chokpak Pass

The Tien Shan mountains are a massive obstacle to migrating birds and few passes cut through them. That fact of geography explains the importance of Chokpak Pass, in the Talasskiy Alatau foothills west of Taraz, which operates as a portal connecting north and south. While mountains on either side tower up to 2,900 metres, the narrow pass is a mere 1,200 metres and connects the vast steppes of Kazakhstan and (beyond those) the forests of the West Siberian Plain to the north with the lowlands to the south. It is

little wonder that this is where migration is most concentrated as birds pass en route to breeding grounds in Central Asia and Siberia in spring, or wintering domiciles in South Asia, the Middle East and Africa in autumn. Millions of birds pass through here in both seasons; those caught in the Heligoland traps provide valuable information for teams of ringers.

Sandgrouse, larks and finches

Spring migration begins at the pass at the end of February. From then through March many Black-bellied Sandgrouse, Skylarks, Calandra and Bimaculated Larks, Rock Pipits, White and Masked Wagtails, Rock Sparrows, Common Starlings, Chaffinches and Bramblings pass through. In April and May there are good numbers of Turtle and Rufous Turtle Doves, European Bee-eaters, Barn Swallows, Sand Martins, Tree Pipits, Yellow, White and Citrine Wagtails, Rosy Starlings, Spanish Sparrows, Ortolan and Red-headed Buntings. Later migrants include European Nightjars, European Rollers and Golden Orioles.

Easterly headwinds usually produce the best results, with scarce and rare birds often found in the Heligoland traps. Corncrake, Common Kingfisher, Striated Scops Owl, Rufous Bush Robin, Eversmann's Redstart, Upcher's Warbler, Asian Paradise-Flycatcher, Long-tailed Shrike and Crimon-winged Finch are some of the more unusual species to show up. Westerlies with

Tired Common Cuckoos (*Cuculus canorus*) on passage can pitch down even on tiny bushes in the Taukum Desert. The species appears in Kazakhstan in April and May and departs in August and September.

169

cloud are unproductive, however. The weather is notoriously variable here, and there can be snow and sub-zero temperatures even in early May. Ringers have described the incongruity of watching European Rollers fly over a snow-covered landscape.

Raptors, cranes ... and more raptors

While Chokpak Pass is good for passerines and near-passerines in both spring and autumn, for cranes and raptors it is much better in autumn although fewer waders use it at that season. Birds of prey have been logged passing at a rate of 2,000 per hour in mid-September. While European Honey-buzzards peak earlier, the third week of September is good for raptor watching. More than 3,000 Steppe Buzzards can pass in a single day, with hundreds of Eurasian Hobbies and Lesser Kestrels and a good mix of other raptors: Oriental Honey-buzzard (at the most westerly extremity of its migration route), Osprey, Short-toed Eagle, Montagu's, Pallid and Western Marsh Harriers, Eurasian Sparrowhawk, Goshawk, Booted Eagle, Lesser Kestrel and Red-footed Falcon, for example. The best time for European Bee-eaters is early to mid-September. Yellow-eyed Stock Doves and Oriental Turtle Doves pass in late September and October.

As with any bottleneck site, visible migration can be spectacular on some days and almost non-existent on others. Birds that pass in large numbers are European Bee-eater (1,000 a day is not unknown, and more than 100,000 were counted in autumn 1999), Greater Short-toed Lark, Sand Martin, Barn and Red-rumped Swallows, Tawny and Red-throated Pipits, warblers, corvids (more than 100,000, mostly Rooks, were once logged in a two-hour period), finches (including Common Rosefinch) and Red-headed Bunting. Laughing Dove, Nightjar, European Roller, Northern and Pied Wheatears, Black-throated Thrush, Greenish and Hume's Warblers, Bearded Tit, Isabelline and Lesser Grey Shrikes and Common Myna are just a few of the many other regulars. Shikra and Merlin are scarce.

Korgalzhyn State Nature Reserve

Far to the north of Chokpak and south-west of the capital, Astana, the enormous lake complex of Tengiz-Korgalzhyn could not be more different from the pass. The lakes sit just 300 metres above sea level in the midst of gently rolling countryside, which is an icy waste in winter and baked by scorching sunshine in summer. Their combined area is about 200,000 hectares, the size of West Yorkshire. Salty Tengiz Lake is the largest, and to the east are the Korgalzhyn lakes. All these lakes are very shallow, and most (though not Tengiz) have extensive reedbeds. Particularly in spring they service mind-boggling numbers of birds en route to the Arctic. The most northerly breeding Greater Flamingos (up to 14,000 pairs) use mud islands on Tengiz Lake, while up to 4,000 pairs of Dalmatian Pelicans nest in the vast reedbeds of the Korgalzhyn lakes.

Red-crested Pochard (*Netta rufina*) are later arrivals in spring than some other wildfowl, mostly in March and April. They migrate in pairs or small flocks of up to a couple of dozen. After the breeding season, large numbers congregate on lakes in late June and July to moult before moving south again in September and October.

However, these wetlands' value as a stopover cannot be underestimated. Hundreds of thousands of Greater White-fronted Geese, Common Pochard, Common Coot and Red-necked Phalaropes pass through, along with thousands or tens of thousands of Greylag Geese, Ruddy and Common Shelduck, Gadwall, Eurasian Wigeon, Mallard, Northern Shoveler, Northern Pintail, Garganey, Eurasian Teal, Tufted Duck, Common Goldeneye, White-headed Ducks, Great Crested Grebes, Demoiselle and Common Cranes, Black-tailed Godwits, Spotted Redshank, Temminck's and Little Stints, Curlew Sandpipers, Ruff and Pallas's Gulls. Smaller, but still significant, numbers of Lesser White-fronted Geese (970 in October 1999 was exceptional), Red-breasted Geese (in October these appear from the north, rest and feed, then fly west towards the Caspian Sea) and Sociable Lapwings are some of the other regulars. The last is usually seen in only small numbers but in August 2011 there was a flock of 500–650 in this area, at Arykty. Migrant raptors, larks, pipits, wagtails and other passerines can be seen more or less anywhere at the right times of year.

The Taukum Desert

This vast area, between Almati and Lake Balkhash, is good in spring when there is more water. Small pools with sparse vegetation around them are

A stunning male Rufous-tailed Rock Thrush (*Monticola saxatilis*). The first birds of spring return to mountain slopes and – on passage – steppes and desert plains in early April. Return migration takes place in August and the first half of September.

magnets for passage birds and newly returned breeding species alike. The latter include exciting species such as Caspian Plover, Greater Sand Plover, MacQueen's Bustards (which winter in Iran and southern Afghanistan), Pallas's Sandgrouse, White-winged Larks, Rosy Starlings (it is possible to see thousands in a day), Sykes's and Desert Warblers, and Mongolian and Desert Finches. It is often hard to tell which are passing through and which are staying. These oases are worth checking for waders (including Temminck's Stint, Red-necked Phalarope and Collared Pratincole), half a dozen species of larks, several races of 'Yellow' Wagtails, pipits, chats, a plethora of warbler species, shrikes, finches and buntings. And where there is prey, there are also raptors. The area around Konshengel is good in April and May.

Sorbulak Lake

Sorbulak Lake and a system of large and small sewage reservoirs lie in the desert north-east of Almaty. Although all the lakes are fresh water, the smaller of them become brackish in late summer and autumn. Sorbulak Lake is 35 kilometres long and is important for breeding, wintering and passage waterbirds. The last include up to 50,000 Mallard, 35,000 Northern Pintail, 15,000 Red-crested Pochard and 40,000 Common Coot. In autumn, big flocks of Ruddy Shelduck and Smew congregate, with a peak of 12,000 of the latter. Good numbers of White-tailed Eagles also arrive in autumn. Tens of thousands of waders (including Terek and Curlew Sandpipers, Common Greenshank, Spotted Redshank, Ruff and Collared Pratincole) use the lake margins and passerine migrants – wagtails, pipits, Common and Rosy Starlings, Barn Swallows and Eurasian Crag Martins – feed and roost in the waterside vegetation. Rare vagrants sometimes turn up.

Sociable Lapwing

Listed by BirdLife International as Critically Endangered, this bird of the Kazakh and south-central Russian steppes suffered a major contraction in its breeding range during the late 20th century. In its core range, north and central Kazakhstan, its numbers fell by 40 per cent between 1960 and 1987. However, a 2006 census found 376 breeding pairs. When extrapolated over the whole range this produced an estimated 5,600 breeding pairs, but the conversion of steppe to arable farmland, and illegal hunting, threaten its survival.

Sociable Lapwings form flocks at the end of the breeding season, before they migrate across southern and western Kazakhstan to winter quarters in the Middle East and India. In August 2011 a flock of 500–650 adults and fledged young was seen near Korgalzhyn – the biggest flock in Kazakhstan since 1939. Birds migrate to the south-west in parties of 15–20 between mid-August and early October. In October 2011 a record flock of 3,200 was found in Turkey; it included a satellite-tagged bird from Kazakhstan. Satellite tagging has also shed light on previously unknown wintering areas, including Sudan. In spring, Sociable Lapwings start their migration in March and early April, reaching Kazakh breeding territories between mid-March and early May. The chances of finding a passage bird are very slim but there are autumn records from Chokpak Pass.

PAKISTAN
INDIA
● GREAT RANN
Bhuj ●
● LITTLE
RANN
Gulf of Kutch
GUJARAT
Arabian Sea

- **LOCATION**

  Gujarat, India

- **FLYWAY**

  Central Asian

- **AUTUMN**

  Wildfowl, including Northern Shoveler, Garganey, Eurasian Teal, Eurasian Wigeon, Northern Pintail and Pochard; Great White and Dalmatian Pelicans; storks; many raptors; Common and Demoiselle Cranes, hundreds of thousands of waders; larks, pipits, wagtails, warblers and other passerines including Grey Hypocolius and Rosy Starling.

- **KEY SITES**

  Great Rann, Little Rann, Chhari Dhand wetland, Banni grassland, Nawa Talao Lake, Nalsarovar Lake, Thol Sanctuary, Velavadar.

- **THREATS**

  Overgrazing of grasslands.

Rann of Kutch, India

The Rann of Kutch in Gujarat, India, is the world's largest salt desert. It is a desolate place in the dry season but after the summer monsoon rains it is invaded by hundreds of thousands of waterbirds, producing some of the most spectacular birding sights on the planet.

Before the monsoon, in May, the Rann is one of the most inhospitable places imaginable, a huge salt-impregnated flatland covering thousands of square kilometres, with a few islands of vegetation rising above it. However, with the onset of the summer monsoon, everything is transformed. The rains start in June and continue until September. By July the rationale for the Rann's designation as a Ramsar site becomes clear. Seawater inundates the low plain from the Arabian Sea (of which it was once a part), and fresh water from monsoon-swollen rivers floods huge areas with brackish water up to 2 metres deep. The region is split into two: the Great Rann along the border with Pakistan and north of Bhuj, and the Little Rann between Bhuj and Ahmedabad to the south-east. Within each are vegetated islands called *bets*, some of which are large. For example, the Pung Bet in the Little Rann is 30 square kilometres in area and supports India's only colony of breeding Lesser Flamingos.

The invasion

With the water come the birds: wildfowl from the Siberian tundra; flamingos and Great White Pelicans from other parts of India; Dalmatian Pelicans and storks from Russia, Kazakhstan and the Middle East; Common Cranes from northern taiga forests; and Demoiselle Cranes, Sociable Lapwings, larks and Rosy Starlings from Central Asian grasslands. Other species, such as Sarus Crane, are only short-distance migrants, coming to where the water is. The flamingos come to breed, but most of the other species have

Numbers of Black-eared Kites (*Milvus lineatus*) increase in the Rann of Kutch in winter as birds move into the area from higher ground to the north.

Sarus Crane (*Grus antigone*) is subject to local movements, abandoning the Rann of Kutch in the dry season for well-watered country, marshes and rivers elsewhere.

already finished their breeding season. Many of the arrivals use the wetlands as a staging post for destinations further south-east, but almost incalculable numbers remain.

In winter the grasslands around the ephemeral lakes welcome thousands of raptors, including the world's largest harrier roost. Greater Spotted, Eastern Imperial and Steppe Eagles join resident Tawny and Bonelli's Eagles. Common Cranes, MacQueen's Bustards, Spotted Sandgrouse, Greater Short-toed and Bimaculated Larks, Desert Warblers, Rosy Starlings and many other passerines also take up residence. Some of the few sites to see Lesser Florican are nearby, as are India's only wintering Grey Hypocolius. Birding interest peaks between mid-November and mid-April, after which the water has drained away. The area supports at least 370 species of birds, some of which are wintering species while others are breeders.

A seasonal summary

Flamingos congregate at colonies from May onwards, just as the monsoon rains begin. From July to September endangered Lesser Floricans display around Surendranagar, just south of the Little Rann. In September and October straggling lines of Common Cranes cross the border from Pakistan to take up their winter domicile around the region's many wetlands. Hundreds of thousands of wildfowl follow them, targetting lakes such as Nalsarovar, Velavadar and Thol, as well as the Great Rann and Little Rann. Waterbird numbers have peaked by December, when they are constantly bothered by eagles and harriers. The Velavadar harrier roost is at its breathtaking best at this time and during January. As water levels recede in February and March, the winter visitors leave the area, moving to those pools that still hold water, the coast or breeding grounds to the north. The timing of spring migration depends on how intense the monsoon rains were and how long the area can keep its water.

Great White Pelicans (*Pelecanus onocrotalus*) rest between fishing expeditions. As the monsoon rains draw to a close, large numbers of pelicans descend on wetlands such as Nalsarovar Lake and the Great and Little Rann.

176

The Great Rann

Together with its Little neighbour, the Great Rann covers about 30,000 square kilometres – considerably larger than Wales. More than a dozen *bets* stand out above its waters when it is inundated, including those of Kakida, Sol, Tragadi, Khadir and Pachchham. One of our planet's biggest concentrations of birds winters on and around the Great Rann, which stretches from Vav in the east to Lakhpat in the west. South of the seasonally flooded area is the grassland of Banni, which does not rise above 5 metres above sea level. Merlins and Short-eared Owls hunt over the grassland, on which MacQueen's Bustards and large flocks of Greater Short-toed Larks feed. Near the village of Fulay is the seasonal wetland of Chhari Dhand, which gets swampy only after monsoon rains. This is another spectacular location when the autumn migrants arrive. It attracts 50,000 or more waterbirds, including huge numbers of wildfowl and 30,000 Common Cranes, as well as many birds of prey. The cranes stream across the border from Pakistan in the second or third week of September, having flown down the Indus Valley. Many remain at Chhari Dhand through the winter, while others move on to different seasonal wetlands.

Wet post-monsoon wetlands near Fulay provide feeding for Red-throated and Tree Pipits, and in February 2007 there was a flock of 60 Water Pipits, previously unknown in Gujarat. Before it dries out in spring, the shallow saline lake of Hodko Dhand may have hundreds of feeding Lesser Sand and Kentish Plovers, Little Stint, Ruff and smaller numbers of other waders. Gujarat's first Caspian Plover was here in February 2007.

The Little Rann

The Little Rann is the southward extension of the Great Rann and is easier to work. Dasada makes a good base, particularly as the village is in a good area for Sociable Lapwings (there were 45 in November 2007) and Rosy Starlings in winter. Nearby, Nawa Talao Lake holds good numbers of wildfowl, including Garganey, as well as Great White, Dalmatian and Spot-billed Pelicans; Glossy, Black and Black-necked Ibis; assorted egrets; Greater and Lesser Flamingos; six species of storks; and a variety of waders. Tracks lead around the Little Rann, providing great opportunities to witness the grand spectacle.

Nalsarovar Lake

This is a large freshwater lake about 40 kilometres south of Dasada and 80 kilometres south-west of Ahmedabad. It expands to 120 square kilometres in the wet season, with its waters surrounding hundreds of islands. As the monsoon rains draw to a close, large flocks of Common and Demoiselle Cranes, Greater and Lesser Flamingos, Great White and Dalmatian Pelicans, storks, wildfowl and waders start to descend on this wetland, reaching a peak in midwinter. Co-ordinated counts here and at Thol in January or February produce figures of between 170,000 (2012) and 250,000 (2008) migrant waterbirds, and these exclude birds that use smaller local wetlands such as

Bhaskarpura and Vadla. Line upon line of Glossy Ibis can be seen flying over Nalsarovar before dawn; there are thousands of Northern Shoveler, Garganey, Eurasian Teal, Eurasian Wigeon and Northern Pintail; tens of thousands of Sand Martins feed around stands of reeds; and buffaloes disturb thousands-strong flocks of waders. Among all this frantic activity there are small numbers of Sarus Cranes and Pallas's and Little Gulls. The lake is at its most dramatic between November and February. A few kilometres away, midway between Ahmedabad and Dasada, the Thol Sanctuary is mostly open water surrounded by marshes and scrub forest. In winter it holds Great White Pelicans, Greater and Lesser Flamingos, cranes (including Sarus) and wildfowl. The biggest numbers are here from November to February.

The Velavadar harrier roost

South again, a little north of Bhavnagar, is Velavadar, where breeding Lesser Floricans display during the monsoon and there is a resident population of the limited-range Stoliczka's Bushchat. One of the site's other avian claims to fame is its winter harrier roost. Harriers arrive in September and their numbers build through the autumn to peak in December, when more than 2,000 have been counted. Most are Montagu's, but there are also large numbers of Pallid and a small representation of Western Marsh and Hen Harriers. They pour into the roost just after sunset, when – with a bit of luck – it is possible to see 500 in the air at one time.

Grey Hypocolius

A local speciality in the Rann of Kutch is the wintering flock of Grey Hypocolius, the only one in India and the most easterly in the world. Up to 150 of these birds arrive in scrub forest near the village of Fulay (south of the Great Rann) in October or November and remain until March or April. This species is the sole member of the family Hypocoliidae. It is a bird of semi-arid and arid countryside and cultivated land. It breeds in an arc from northern Iraq to southern Afghanistan and winters further south, from Saudi Arabia and Yemen to Gujarat.

In winter the grazing marshes running along the Gulf of Khambhat hold Eurasian Teal, Northern Pintail, Common Pochard, Glossy and Black Ibis, Greater and Lesser Flamingos, Little and Temminck's Stints, Wood and Green Sandpipers and Citrine Wagtails. Visit in August for the floricans or between December and March for the waterbirds and raptors.

Gir Forest

South of Visavadar, Gir Forest National Park and Wildlife Sanctuary comprises about 1,500 square kilometres of dry deciduous forest (including teak), scrub forest and dry grassland. There are also reservoirs, drinking pools for the area's large mammal population and several perennial rivers. The park is most famous for having the world's only population of Asiatic Lions, which numbered more than 500 in 2011. There are several other scarce mammals, including Indian Leopard.

Many northern-breeding birds arrive in autumn. Waders include Jack Snipe, Marsh Sandpiper, Temminck's Stint and Ruff, and wintering raptors include Western Marsh, Hen, Pallid and Montagu's Harriers. Tawny and Blyth's Pipits, Yellow, Citrine and White Wagtails, several species of wheatears, Paddyfield, Sykes's, Eastern Orphean and Greenish Warblers, Lesser Whitethroat, Common Chiffchaff and Golden Oriole are among the passerine migrants that visit for the months of winter.

Varnu Temple near Adesar, at the western edge of the Little Rann of Kutch. The arid plain beyond is inundated by August or September but drains again by the following April.

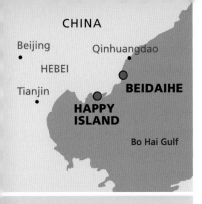

CHINA

Beijing
· HEBEI
Qinhuangdao
·

BEIDAIHE
○

Tianjin
·
○
HAPPY
ISLAND

Bo Hai Gulf

- **LOCATION**
  North coast of Bo Hai Gulf, Hebei, China

- **FLYWAY**
  East Asian–Australasian

- **SPRING**
  Wildfowl, buntings and cranes pass through from early March. Shorebirds follow, then activity builds to a climax in May, with more buntings, wildfowl, herons, raptors, crakes, more waders, cuckoos, needletails, hirundines, chats, thrushes, warblers, and white-eyes. Passage of scarcer Locustella warblers runs until early June.

- **AUTUMN**
  Return passage starts in late June but is only really marked from late August. Late-arriving spring migrants are often early returners. Waders move through from July, followed by large numbers of passerines, raptors, Oriental White Storks and, in November, five regular crane species.

- **KEY SITES**
  Lighthouse Point, Sandflats and Lotus Hills, Beidaihe; Big Wood; Liaojian Harbour; Happy Island.

- **THREATS**
  Development pressure all along this stretch of coast.

Beidaihe, China

The coast of the Bo Hai Gulf, a few hours' drive east of Beijing, is one of the most exciting places on Earth to experience the full spectacle of migration.

At peak season, in mid-May, anywhere between the muddy fringes of Liaojian Harbour to the mouth of the Stone River – sites centred on world-famous Beidaihe – can be the recipient of a huge early morning fall of chats, thrushes and warblers. Later in the day the same location may bear witness to a spine-tingling movement of raptors. Alternatively, large numbers of northward-bound waders may appear on the coastal mud as the tide ebbs. The lucky birder, and I've been in that position, will have all three experiences in the same day.

Beidaihe was once the favoured holiday destination of Chairman Mao. Today it is one of the very best places to taste the exotic flavours of the great East Asia Flyway, in spring and autumn. This conveyor belt carries birds from and to Australasia, South-east Asia, India and even Africa in spring and autumn as they move between wintering and breeding quarters. The numbers of birds involved are mind-blowing. Of course, only a minority are ever seen by human eyes but if conditions are good one can imagine the true scale of this avian mass movement, as anyone who has heard or seen 200 Radde's Warblers in one small area (before abandoning the count) will testify.

The year in perspective

While May and the middle two weeks of October are the periods with the greatest variety, migration touches the area from

This male Blue Rock Thrush (*Monticola solitarius*) of the eastern race *philippensis* had just arrived at Lighthouse Point, Beidaihe.

March to November. As the last *Acrocephalus* stragglers are making their way north in June, the first returning terns and waders may already be heading in the other direction. Only the frigid months of December, January and February are without interest: Beidaihe supports just a handful of species at this time.

In March, Bean Geese, Common Goldeneye, Common, Red-crowned, Siberian and Hooded Cranes, Northern Lapwings, Siberian Accentors, Eurasian Skylarks, Rooks, Daurian Jackdaws and Rustic Buntings will already be on the move. Things pick up in April, when wildfowl figure increasingly: Falcated, Spot-billed and Mandarin Ducks and Garganey pass through, and there is the chance of the endangered Baer's Pochard. Passerines such as Red-flanked Bluetails, Pallas's Warblers, Yellow-throated Buntings and Long-tailed Rosefinches can also be found. In the last week of April, the steady flow becomes a rush, and a two-week slot around the middle of the month produces more than 200 species of migrants pretty much anywhere along this stretch of coast. Wildfowl are not now so prominent, but most other groups are. Raptors may appear in good numbers: Oriental Honey-buzzards, Eastern Marsh and Pied Harriers, Japanese, Eurasian and Chinese Sparrowhawks, Grey-faced and Upland Buzzards, Common Kestrels, Amur Falcons, Eurasian Hobbies, Sakers and Peregrines all put in appearances, some in big numbers,

Yellow-rumped Flycatcher (*Ficedula zanthopygia*) winters in peninsular Southeast Asia and breeds in mature forest in north-east China. Bright males like this bird can be expected at Beidaihe in May, even in the grounds of the town's hotels.

others not. Waders stream north along the coast, stopping to rest and feed at shrimp pools or on mudflats. Their ranks include Kentish Plovers, Lesser and Greater Sand Plovers, Pacific Golden Plovers, Grey-headed Lapwings, Great and Red Knot, Red-necked and Temminck's Stints, Sharp-tailed, Curlew and Broad-billed Sandpipers, godwits, Far-eastern Curlews, Marsh, Wood, Green and Terek Sandpipers and many more. Among the commoner species will be a few of the real gems: Oriental Plover, Spoon-billed Sandpiper, Nordmann's Greenshank, Asian Dowitcher and Grey-tailed Tattler.

Chats, thrushes and flycatchers

In May, migrant passerines will be in areas of coastal cover and especially – it would seem – in the grounds of Beidaihe's hotels. The thrushes that a visitor flushes from a hotel lawn may be an adrenaline-pumping mixture of White's, Siberian, Grey-sided, Grey-backed, Pale, Eye-browed and Dusky. A single flowerbed may conceal Siberian Rubythroat, Bluethroat, Siberian Blue Robin and a late Red-flanked Bluetail. Small shrubs provide hawking perches for Asian Brown, Taiga, Mugimaki and Yellow-rumped Flycatchers. Then there are the warblers and buntings. Some of the latter have already passed through by May, but there are still Black-faced, Tristram's, Little, Chestnut and Yellow-breasted to find. Coastal reeds will sometimes attract Pallas's Reed and Japanese Reed Buntings. For those who like their warblers, Beidaihe is hard to beat. Phylloscs begin to arrive in April and numbers build in early May, peaking around the middle of the month. Towards the end, it is the turn of the *Locustella* and *Acrocephalus* warblers to stream though, sometimes appearing in the most

The view from Beidaihe Sandflats towards Lighthouse Point. In autumn, the Sandflats are a great place from where to watch overflying cranes. In spring they can be good for migrant waders and for watching visible migration.

for Lanceolated Warblers. Mind you, these can and do turn up anywhere. Maybe 400 metres from Suzy Wong's are the grounds of the Friendship Hotel. Manicured lawns, some slightly rougher grass, flower beds, and shade trees hardly seem to add up to a migration hotspot, but appearances can be deceptive. This place is often where migrants end up after they've filtered through from initial landfall at Lighthouse Point. Pride of place here goes to the thrushes. In May, Siberian, Grey-sided, Grey-backed, Eye-browed, Dusky and Chinese Song Thrushes are all likely to be seen, sometimes feeding together on the grass, with Eye-browed and Dusky regular and in good numbers. This is also a great place for dazzling views of kaleidoscopic Mugimaki and Yellow-rumped Flycatchers. A Forest Wagtail may fly overhead, giving away its presence with a 'pink-pink' call.

The Sandflats

On the northern side of the town is Pigeon Nest Park and Sandflats, which is now a no-access area of beach, thanks to the campaigning work of conservationists. An extensive area of sand is exposed at low tide. The area can be viewed from a walkway, which is raised in part. The Sandflats are attractive to herons, cranes, waders and sometimes Mongolian Lark at the appropriate times of year. There is also a small reedbed, good for Pallas's Reed and Japanese Reed Buntings. In spring, look for Kentish and Lesser Sand Plovers, Red-necked Stints and Little Whimbrel here.

Over the bridge from the Sandflats is Yeng Ho Reservoir, now called a 'wildlife park'. The reservoir itself is bordered in places by reedbeds and has reedy islands. To the north are some open-canopy woodland, small pools and reedbeds. As well as Great, Yellow and Schrenk's Bitterns and Purple Herons, the reedbeds are also good for 'acros', and the woodland attracts a variety of migrating warblers, chats, thrushes and flycatchers.

Lotus Hill, Stone River and Yang He

The elevated position of the Lotus Hills (Lianfengshan), just to the west of the town and standing 150 metres above sea level, makes them a good location for visible-migration watching, especially for raptors, cranes, needletails and hirundines. On 2 November 1990, after a switch in wind from north-easterly to southerly, 389 Siberian Cranes were logged passing the hills, among Red-crowned, Common and Hooded Cranes. More recently, one lucky group of birders counted more than 1,000 cranes of seven species pass over in 90 minutes! These numbers were exceptional but indicate what is possible.

Half an hour's drive north-east of Beidaihe, and relatively unknown, is the estuary of the Stone River, which at low tide has good patches of exposed silt and shallow pools. These – and the offshore bars – are worth checking for Red-necked Stints, Sharp-tailed Sandpipers, *Tringa* sandpipers, Terek Sandpipers and Grey-tailed Tattlers. Large flocks of Common Whimbrel may gather, and it is also very good for Little Whimbrel. Baillon's Crakes use the shrimp pools on passage. The grassland and stunted trees running along the coast here attract pipits and flycatchers.

Travelling south-west, the Yange He shrimp pools provide a great opportunity to get close views of passage waders, but you have to find the right pools. Some are too dry, others too full of water, but those with just a little water become magnets for Temminck's and Long-toed Stints, Marsh, Sharp-tailed and Curlew Sandpipers, Far-eastern Curlews, Eastern Black-tailed Godwits and more. Walk along the bunds between the pool basins, taking care

The *florensis* race of Brown Hawk Owl (*Ninox scutulata*) is a summer visitor to north-east China. On migration it turns up along the coast near Beidaihe, typically in the middle of May.

Asian Dowitcher (*Limnodromus semipalmatus*) is classified as Near Threatened, and its Chinese breeding population is probably fewer than 100 pairs. It winters in South-east Asia, using Mai Po and the coast around Happy Island (23 birds together in May 2012) as staging areas.

not to spook any birds in them. This area can also be good for pipits, Daurian Starlings and buntings. After a fall, Chestnut-eared, Yellow-browed, Tristram's, Chestnut, Yellow-breasted, Pallas's Reed and Japanese Reed Buntings may be found. Siberian Stonechats sometimes number scores, along with Chinese Penduline Tits.

Magic Wood and Liaojian Harbour

The grounds of Beidaihe's Friendship Hotel sometimes attract passage Siberian Thrushes (*Zoothera sibirica*) but Magic Wood and Happy Island are probably better bets for the species, which is more of a skulker than most other migrant thrushes.

Lying a three-hour, 80-kilometre drive south-west of Beidaihe is a cluster of sites, some of whose names hint at birders' experiences of former years: Big Wood, Magic Wood, Liaojian Harbour and Happy Island. This is an area where development is taking place at breakneck speed. Coastal marshes are being transformed by industrial, residential and leisure development on a scale that is difficult to comprehend. It's anybody's guess how long these places will remain 'magic' and 'happy' but while there is some cover I suspect that tired birds making landfall will continue to drop in.

Big Wood is not as big as it once was but can still attract any number of passerines. Magic Wood, slightly closer to the coast, is tiny – but incredible. Part of it houses a Black-crowned Night Heron colony, which has moved from Big Wood; there is no access to this section. The other half of the wood is a small rectangle with some quite thick cover. One day it may hold 200 or more Yellow-browed and Radde's Warblers, with a similar number of Dusky Warblers a couple of days later. It is close enough to the coast to offer grandstand views of passing waders. Meanwhile, raptor migration can be spectacular: in May 2011, 113 Amur Falcons and 24 Eurasian Hobbies passed overhead in just a few hours. The same afternoon, two Black-faced Spoonbills flew over. Anything from Yellow-legged Buttonquail and *florensis* Brown Hawk Owl to Pied Kingfisher can turn up here. Nearby, Liaojian Harbour can be very good in spring and autumn if the tide is right: low but rising. Relict and Saunders' Gulls are

regular, and one day in mid-May 2011, 1,200 Great Knot were strung out along the tideline – an awesome sight.

Happy Island

Any birder who has visited the island at the height of spring migration will agree that it was not misnamed. Just a few minutes' motorboat ride offshore, its classic spring falls are the stuff of legend. Only 4 kilometres by 2.5 kilometres, despite development in recent years there is still plenty of cover to encourage migrants to stay awhile. Birders have experienced some of the most dramatic falls on Happy Island, and the surrounding mudflats can hold thousands of waders. It is internationally important as a staging post for numerous threatened species, especially Great Knot and Asian Dowitcher (23 together in May 2012) and the occasional Spoon-billed Sandpiper, with Spotted Greenshank in autumn. More expected waders include Dunlin, Black-tailed and Bar-tailed Godwits, Eurasian Curlew, Spotted Redshank and Terek Sandpiper. In May 2012 there were an impressive 237 of the last species.

The woods are a fantastic migrant trap and regularly get an array of vagrants that have become caught up in the main migration. Exceptional birds over the years have included Fairy Pitta, Japanese Robin, various *Seicercus* warblers and a flock of 75 Siberian Thrushes. The newly built golf course should attract lots of pipits and is potentially good for Little Whimbrel and perhaps the odd Oriental Plover. The island is also notable for spring cuckoos, Grey Nightjar, *Phylloscopus* warblers, Black-naped Oriole and buntings.

In mid-May the first Lanceolated Warblers (*Locustella lanceolata*) begin to arrive at Beidaihe en route to wet grasslands in Siberia. The species' mouse-like behaviour means it can 'disappear' in even tiny areas of grass, but migrants can sometimes be showy.

189

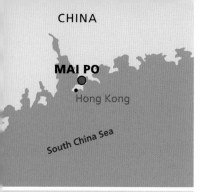

- **LOCATION**
  Hong Kong, China

- **FLYWAY**
  East Asian–Australasian

- **SPRING**
  Wader passage often
  begins with Curlew
  Sandpiper in early March,
  then builds. Mid-April to
  early May is the most
  exciting period with
  hundreds or thousands
  of Black-tailed Godwit,
  Common Greenshank,
  Marsh Sandpiper, Terek
  Sandpiper, Red-necked
  Stint, Pacific Golden
  Plover and other waders;
  among them may be a few
  Spotted Greenshank, Asian
  Dowitcher and Spoon-billed
  Sandpiper; Red-necked
  Phalarope, terns and skuas
  pass to the south of Hong
  Kong Island.

- **AUTUMN**
  Wildfowl, Black-faced
  Spoonbill, waders and
  Saunders' Gull return from
  September to November.

- **KEY SITES**
  Mai Mo WWF reserve,
  Tai Po Kau, Po Toi island.

- **THREATS**
  Mai Po is protected, but
  development around Inner
  Deep Bay is a constant
  threat.

Mai Po, Hong Kong

Imagine the scene. It is April, and you are sitting in a hide overlooking the mudflats at Hong Kong's Inner Deep Bay. Hundreds of Curlew Sandpipers and other waders are coming ever closer, driven by the rising tide. You scrutinise every one. In fact, you didn't realise you could concentrate so hard. All of a sudden you see what you've been looking for: one small wader with a spoon-shaped bill.

The Mai Po wetland borders Inner Deep Bay, in the north-west part of Hong Kong's New Territories. Its reputation was built on the availability of its 'big three' waders – Spoon-billed Sandpiper, Spotted (formerly Nordmann's) Greenshank and Asian Dowitcher – on spring passage. All three have suffered big declines in global populations, but it is still possible to connect with them here (though the first is far from guaranteed). And Mai Po has a whole lot more to offer. In spring and autumn 20,000–30,000 shorebirds use the pools of Mai Po and the mudflats of adjacent Deep Bay to refuel and rest. More than 50,000 waders, wildfowl, herons and other waterbirds, including 20 per cent of the world's Black-faced Spoonbills, spend the winter there as does a significant proportion of the planet's Saunders' Gulls. In total, more than 380 species have been recorded. Mai Po is so good because it offers food and places to rest – and is firmly on the East Asian–Australasian Flyway, and this value is enshrined in Inner Deep Bay's status as a Ramsar wetland of international importance and Mai Po's protection as a WWF reserve.

The reserve comprises mangroves, shallow shrimp ponds called *gei wais*, freshwater swamps and reedbeds. A network of paths and boardwalks runs between the *gei wais*, connecting various hides, including several that overlook the mudflats of Inner Deep Bay.

A waderfest

Thousands of waders use the intertidal zone for feeding in autumn, winter and spring. Numbers peak in spring when the commonest species are Curlew Sandpiper (up to 5,000), Common Redshank (2,000), Marsh Sandpiper (1,600), Red-necked Stint (over 1,000), Spotted Redshank (1,000), Black-tailed Godwit (750), Common Greenshank (700), Greater Sand Plover (450), Terek Sandpiper (400) and Pacific Golden Plover (300). Spring is the time when the real megas appear, albeit in much smaller numbers: Oriental Pratincole, Spoon-billed Sandpiper, Asian Dowitcher, Little Whimbrel, Far-eastern Curlew and Spotted Greenshank. There are other regular members of the cast, too: Black-winged Stilt, Kentish Plover, Lesser Sand Plover, Grey Plover, Great Knot, Red Knot, Sanderling, Sharp-tailed Sandpiper, Dunlin, Broad-billed Sandpiper, Bar-tailed Godwit, Whimbrel, Wood Sandpiper, Grey-tailed Tattler, Ruddy Turnstone and Red-necked Phalarope. American species such as Long-billed Dowitcher and Green-winged Teal are not unknown.

The mix is different in autumn. Most species appear in smaller numbers (Red-necked Stint, Curlew Sandpiper, godwits, Marsh Sandpiper, Grey-tailed Tattler, for example) although a few are more common, including Black-winged Stilt. However, without many of the rarities, autumn does not have quite the same appeal.

The 'big four'

For four endangered species with breeding ranges in eastern Siberia, Mai Po–Deep Bay is a critical spring stopover. Asian Dowitcher (Near Threatened) has a global population of around 23,000. In spring 2000, there was a peak of

Curlew Sandpipers (*Calidris ferruginea*) and Greater Sand Plovers (*Charadrius leschenaultii*) take flight at Mai Po. Numbers of these and many other wader species are impressive from March to May.

One of the boardwalks at Mai Po WWF reserve, complete with its own Eastern Great Egret (*Ardea alba modesta*).

57 at Deep Bay. Far-eastern Curlew (Vulnerable) has a global population of just 38,000. Even more rare, there could be as few as 500 Spotted Greenshank (Endangered) and maybe only 330 adults left. With a breeding range confined to a corner of north-east Siberia, this species is in serious decline. In spring 2000 at least 35 stopped off at Mai Po. Then there is the holy grail of wader fans: Spoon-billed Sandpiper. Breeding numbers for this Critically Endangered bird, seemingly poised on the brink of extinction, are now down to 120 pairs. Although still annual at Mai Po, numbers are pitifully low; for example, there were just four in 2012, on 8 March, 25 March and 18 April (two). Black-faced Spoonbill is another species for which Mai Po–Deep Bay is of international importance.

Double-figure counts of Baer's Pochard (30 in January 1987) are sadly a thing of the past, but there was a female of this species on pond 20 (Mai Po) in February 2012. This species, which breeds in north-east China and neighbouring Siberia, is rapidly declining and now considered Critically Endangered. Saunders' Gull (Vulnerable – 2011 population estimate of 22,000–23,000) winters here before migrating to breeding grounds further north in China.

The year in focus

Migration starts in early March when small parties of Curlew Sandpipers will be harbingers of what is to come. Later, hundreds of Black-tailed Godwits, Common Greenshank, Common Redshank and Marsh Sandpipers appear. In

2012 there were single Spoon-billed Sandpipers on 8th and 25th. March is also the best time to connect with a Saunders' Gull. Through April, waders continue to arrive, and depart: large numbers of Curlew Sandpipers, along with their Broad-billed, Terek, Sharp-tailed, Green and Wood cousins, Greater and Lesser Sand Plovers, Pacific Golden Plover, Red-necked Stint and more. There were two Spoon-billed Sandpipers in both April 2011 and April 2012, while Asian Dowitcher, Spotted Greenshank, Far-eastern Curlew and Grey-tailed Tattler are also regular in April. Mid-April to early May is the most exciting period. Chinese (Swinhoe's) Egret, Yellow Bittern and Garganey and passerines will also be passing through. Gull-billed and Caspian Terns will be on the mudflats or on the lagoons when the tide is high. White-winged Black Terns prefer the pools. Early May is just as good, with more than 20 species of waders in a day likely, but things become quieter towards the month's end.

Although waders rightly take centre stage, they are not the only migrant interest. Especially in poor weather, for example if a weather front has brought heavy rain, thousands of House Swifts and Barn Swallows may descend on Mai Po, with smaller numbers of Pacific Swifts, White-throated Needletails, Asian House Martins and Sand Martins. Flycatchers, including Narcissus, *Acrocephalus* warblers and several species of buntings, including Yellow-breasted, can also be seen.

After a midsummer lull, wader return passage is well underway, hundreds of Common Greenshank leading the way. At the end of September or early October the first returning Black-faced Spoonbills appear (November is the peak month; there were 461 in November 2009), followed by Eurasian Teal, Eurasian Wigeon, Northern Pintail and Northern Shoveler.

For wader fans, at least, Spoon-billed Sandpiper (*Eurynorhynchus pygmeus*) is one of the most 'wanted' species. With a tiny – and falling – world population, Mai Po is one of the few places on Earth where this bird is regular.

Working the area

Visit Mai Po on a rising tide (check times at www.hko.gov.hk/tide/cTBTtide.htm). There will be herons and egrets on the freshwater pools at other states of the tide, but as it rises the waders will be pushed off the mudflats of Deep Bay and on to the pools. The best strategy is probably to head first to the hides overlooking the bay, then to check the pools once the tide has peaked. The

hides that overlook the mud are accessed via a boardwalk that passes through mangroves. These hides can get very crowded in autumn, winter and spring so it's best to arrive early. Then work the *gei wai* to the east. Pool 16/17 is often very good for waders. In early May 2012, for example, it held more than 500 waders of 17 species, including Greater Sand Plover, Asian Dowitcher, Spotted Greenshank, Broad-billed Sandpiper, Terek Sandpiper, Red-necked Stint and Greater Painted-snipe. Plenty of information about what birds are present, and where to see them, is available at the visitor centre.

Tai Po Kau

If you need a break from waders, herons and wildfowl, Tai Po Kau nature reserve, 15 kilometres to the east (between Tai Po market and the Chinese University), can be productive. The reserve contains some of the best mature secondary forest in Hong Kong, and this attracts passerine and near-passerine migrants in spring and autumn. These include cuckoos, chats, thrushes, Ferruginous and Narcissus Flycatchers, Chestnut-flanked and Japanese White-eyes, Oriental Dollarbirds, warblers, buntings and other species on migration

Black-faced Spoonbill

Although conservation efforts have been successful in increasing its numbers from a few hundred in the early 1990s to more than 2,000 by 2010, this species is still critically endangered. It breeds in colonies on small islets off the west coast of the Korean peninsula and Liaoning Province in adjacent China. Its wintering range is very dislocated, divided between several countries. Deep Bay is the second most important wintering location, after the Tsengwen estuary in Taiwan. Kyushu and Okinawa in Japan, Cheju in South Korea, and the Red River delta in Vietnam are also important. The first wintering birds arrive at Mai Po–Deep Bay in late September. Some stop off here before moving further south; this explains the biggest count to date, of 461 in November 2009. Others remain all winter, leaving for the trip north to breed in April. Black-faced Spoonbills from Vietnam, Hong Kong and Taiwan follow the coast of China before crossing the southern Yellow Sea to South Korea.

between South-east Asian winter quarters and breeding areas to the north.

Po Toi island

If you fancy a boat trip, it is worth visiting another migrant hotspot – Po Toi, a small island south of Hong Kong Island. Although requiring more effort to reach, the birds listed for Tai Po Kau are generally more plentiful and easier to find on Po Toi. And in spring there is the bonus of Red-necked Phalaropes, terns and other seabirds on the sea crossing. The ferry leaves from Aberdeen, on the south side of Hong Kong Island, and takes 45–50 minutes.

The Hong Kong Birdwatching Society sometimes organises a spring pelagic to search for seabirds. In late April 2012 this resulted in sightings of Japanese Tern, Aleutian Tern, Japanese Murrelet and two Ancient Murrelets. In mid-May of the same year there was further confirmation of the pelagic possibilities when a Red-footed Booby, five Bridled Terns and eight White-winged Black Terns were seen from a cruise ship in one day.

Passerine and near-passerine passage migrants can be seen in Hong Kong as well. One of the most striking is Oriental Dollarbird (*Eurystomus orientalis*).

MYANMAR

Bilauktaung mtns

CHUMPHON ●

THAILAND

Gulf of
Thailand

- **LOCATION**
  Coast of South China Sea, southern Thailand

- **FLYWAY**
  East Asian–Australasian

- **SPRING**
  Japanese Sparrowhawk numbers peak in February. Main raptor passage is between mid-March and early April. Black Baza, Oriental Honey-buzzard, Grey-faced Buzzard and Chinese Sparrowhawk pass in big numbers. Osprey, Crested Serpent-eagle and Jerdon's Baza are possible. Non-raptor migrants include spinetails, swifts, Blue-tailed and Blue-throated Bee-eaters and hirundines.

- **AUTUMN**
  Raptor passage extends from late August to early November, peaking in September and October. The regular species are Chinese and Japanese Sparrowhawks and four other accipiters, Black and Jerdon's Bazas, Oriental Honey-buzzard, Grey-faced Buzzard, Eastern Marsh and Pied Harriers, Crested Serpent and Greater Spotted Eagles and Amur Falcon. Herons, waders, three species of bee-eaters and Black Drongo are also regular.

Chumphon, Thailand

Half a million raptors migrate through the Chumphon area of southern Thailand each autumn, including South-east Asia's biggest counts for many species.

Near the town of Chumphon, in southern Thailand, there are several hill summits – notably Khao Dinsor – where it is possible to eyeball tough-to-see species like Black Baza as they pass a matter of metres away. The place is also good for raptor passage in spring, but it really comes into its own in the autumn. The Chumphon area is so good because of its proximity to the east coast of the narrow Thai peninsula. Birds of prey that breed in South-east Asia, China, Siberia and even Japan pass through here to and from their wintering grounds to the south or, in the case of Amur Falcon, in southern Africa. They are often joined by many non-raptors, including spinetails. They all drift to the coast with the prevailing west to north-west winds.

Migrants are channelled along a 30-kilometre-wide coastal plain, which is conveniently punctuated by steep-sided hills – ideal watchpoints overlooking some nice tracts of forest and, in the distance, the South China Sea.

Spring raptor passage

Spring raptor totals are awesome: more than 50,000 individuals of 15 species were noted moving north in both 2007 and 2008, with the second half of March and early April the best time slot. Japanese Sparrowhawks move north about a month earlier, peaking in February. Grey-faced Buzzards are more numerous in spring

Looking north from Khao Dinsor towards the South China Sea. In autumn, north-westerlies push southbound raptors close to the coast and many drift close to Khao Dinsor.

than autumn, often much more so, presumably indicating that they use different routes in the autumn. From March, large numbers of swifts, Blue-tailed and Blue-throated Bee-eaters and hirundines can also be seen moving north.

Autumn passage

In autumn the totals become simply mind-blowing: in 2011 about 292,000 birds of prey of at least 26 species were logged. This figure, which researchers reckon is probably only half the true number passing through, was made up of more than 124,000 Chinese Sparrowhawks, 103,000 Black Bazas, 36,000 Oriental Honey-buzzards and 10,000 Grey-faced Buzzards. For any birder who has struggled to get brief, distant views of one of these as it passes between two chunks of canopy, Khao Dinsor presents an unrivalled opportunity for observation and photography, sometimes at very close range.

An adult Chinese Sparrowhawk (*Accipiter soloensis*) passes low over Khao Dinsor. In a single autumn, more than 124,000 of this beautiful *Accipiter* have been counted at the raptor watchpoint, with peak numbers in early and mid-October.

First on the agenda, in early September, are Japanese Sparrowhawks. Their early peak period (the second half of September) mirrors their early arrival in spring. Oriental Honey-buzzards, Chinese Sparrowhawks, Peregrines, Eastern Marsh and Pied Harriers, Black Kites and Ospreys are seen in greatest numbers in early or mid-October. These are followed by Grey-faced Buzzards, Shikras, Black and Jerdon's Bazas, Crested Serpent-eagles and Booted and Greater Spotted Eagles. Black Bazas are still on the move well into November. And there is always a rarity waiting to be picked out of the aerial throng: a Besra, Rufous-winged Buzzard, Oriental Hobby or Amur Falcon, for example.

The best weather for viewing in autumn is when winds are from the west, drifting raptors towards the coastal Chumphon ridges. On occasion, the raptor totals then are hard to appreciate. On 23 October 2003, for example, with a westerly breeze and overcast conditions, 56,101 raptors poured through. As the winds shift to the east at the end of October or early November, the flow slows. Raptor flights can still be good in November, but only after rain has

cleared. Jerdon's Baza has been seen in flocks of up to 10 in early November, when Steppe and Eastern Imperial Eagle may also then be possible.

Needletails, bee-eaters and drongos

Apart from the raptors there is plenty of other interest, including the passage of White-throated, Brown-backed and Silver-backed Needletails: about 1,100 were counted in autumn 2011. Also moving through will be swifts and swallows, Blue-tailed, Blue-throated and Chestnut-headed Bee-eaters (27,000 in 2011), Dollarbirds, Ashy Minivets, Purple-backed Starlings and Black Drongos, the last in big numbers. For example, 11,290 were counted between 27 September and 9 November 2003. They leave overnight roosts shortly after dawn, moving south with bee-eaters and swallows, but before the onset of most raptor migration. Herons, Oriental Pratincoles and Grey-headed Lapwings could also feature in the cast of diurnal migrants.

Intensive study at Chumphon has revealed previously unknown nuggets of information. The sheer scale of raptor migration down the peninsula in

Where else in the world could you get such a good view of Black Baza (*Aviceda leuphotes*)? The autumn record count for this species (in 2011) was 103,000, with birds still on the move well into November.

autumn is bigger than once thought: more than 10,000 Black Bazas and over 9,000 Oriental Honey-buzzards in a day, for example. Rarer finds have included a small passage of Jerdon's Bazas (78 in 2011) and Brahminy Kites (18 in 2010, the first suggestion of migratory habits in South-east Asia).

Working the area

In spring, prevailing winds tend to be from the north-east, combining with sea breezes to drift migrating birds of prey several kilometres inland (west). They do not tend to be concentrated in such a dense corridor as in autumn but Khao Promsri, 15 kilometres west of Chumphon city, is the best place to be. Prime times to watch are before 11 a.m. and after 3 p.m. Between these hours, the raptors may rise so high on the massive thermals that they become invisible. In autumn, north-westerlies push raptors very close to the coast, in a concentrated flow. Birders have a choice of summits: Khao Dinsor or Khao Radar. The former involves an uphill climb, steep in parts, but the effort should be rewarded because birds come very close. Khao Dinsor rises 350 metres above the coastal plain, about 25 kilometres north-east of Chumphon.

Conservation issues

The focus on raptor migration at Chumphon is an enormous positive. It was the first Thai city to promote raptor conservation, via roadside signs and even T-shirts. The area's importance for raptor migration is being used to educate local people, a Raptor Education and Research Center opened late in 2012 and local people assist with the counts. The ethos is very much for community based raptor conservation.

Oriental Honey-buzzards

The northern population of Oriental Honey-buzzard (*Pernis ptilorhyncus*) is migratory, breeding in parts of Siberia, China, Korea and Japan, and wintering in South-east Asia. Two adults were tracked by satellite after they left their breeding grounds in Japan. They crossed the East China Sea and passed south through China and the Malay Peninsula to reach Indonesia. The female remained on the island of Java after a 52-day migration, but the male turned north again, passing through Borneo to reach the Philippines in late November, 68 days after setting off and after a trip of more than 11,000 kilometres. Sections of their journeys were over wide expanses of water, and some of their migration was accomplished during the hours of darkness. At the end of February, three months after arriving at their respective winter domiciles, the birds began their return trips.

- **LOCATION**

  In the Titiwangsa Range of central Peninsular Malaysia

- **FLYWAY**

  East Asian–Australasian

- **AUTUMN/WINTER**

  Oriental Honey-buzzard and other raptors; Grey Nightjar; Himalayan Swiftlet and Silver-backed Spinetail; hirundines; Blue-winged and Hooded Pittas; Blue-tailed Bee-eater; Siberian Blue Robin; thrushes, including Eye-browed, Orange-headed and Siberian; warblers; flycatchers; Forest Wagtail; rare 'overshoots' from the north possible.

- **KEY SITES**

  The Gap, Fraser's Hill resort.

- **THREATS**

  Fraser's Hill is protected, but other forest in the range could be threatened by forestry.

Fraser's Hill, Malaysia

Blue-winged Pitta, Blue-tailed Bee-eater, Orange-headed Thrush, Mugimaki Flycatcher and Forest Wagtail are just a few of the beautiful migrants that spend the winter in the forests of this former colonial hill station.

Imagine the scene. It is mid-afternoon, but you are far away from the humid heat of the lowlands. In fact it is pleasantly cool, the low cloud gently wafting around the canopy arching over the trail. It's a while since you've seen a bird-wave but a slight movement catches your eye a little way inside the forest. Bins raised, you focus on where the movement was. Nothing. Then a small bird flits to the ground. Blue and white. There's no doubting the identity of that plumage: it belongs to a glorious male Siberian Blue Robin!

Unlike some migration hotspots, Fraser's Hill offers exciting birds at all times of year. Its resident species include a mouth-watering array of Oriental raptors, partridges, trogons, kingfishers, bee-eaters, hornbills, barbets, woodpeckers, broadbills, flycatchers, bulbuls and babblers. Mountain Peacock-pheasant and Malayan Whistling-thrush are endemic to Peninsular Malaysia. From late October or November their ranks are added to by visitors from the north. Visible migration is not spectacular, though some raptors, swiftlets and spinetails, bee-eaters, hirundines and small flocks of thrushes can be seen flying over. More subtle is the appearance on the forest floor or in the canopy of pittas, bee-eaters, thrushes, flycatchers, phylloscs and more. One day they aren't there, the next they are. Many of these birds' migration patterns are not fully understood, but clearly the forested highlands along the spine of Malaysia form a conduit for millions of passerines and near-passerines in autumn and spring.

Fraser's Hill (or Bukit Fraser), in the Central Titiwangsa Range, was opened up as recently as the 1920s when the first decent road was carved through the forest and a hill station was established. Today, the place has the look of an ex-British colonial settlement, complete with clock tower and post office, though there is also a mosque. The upland rainforest ranges from 800 metres at The Gap to 1,500 metres at Pine Tree Hill and is home – for at least part of the year – to more than 270 species. It has a long history of bird observation and ringing and each June hosts an international bird race.

Nocturnal migrants

Between 1964 and 1988 bird trapping and ringing was conducted in a systematic fashion. Spotlights and mist-nets were used in an attempt to quantify nocturnal migration. The results were fascinating; for example, Tufted Ducks and Masked Finfoots were caught, demonstrating that both migrate at night and at altitude. Blue-winged Pittas were found to make up the majority of migrating pittas during October, but in the following month Hooded Pittas were ascendant; between 1965 and 1973 more than 1,200 of the latter were trapped. The programme also found rarities, including Besra, an accipiter which is resident to the north and south but not in central Malaysia; Siberian Rubythroat, which usually winters no further south than northern Thailand; Alström's Warbler, which has winter quarters as far south as southern Thailand, but not in central Malaysia; and – most intriguing of all – Rufous-headed Robin, which breeds in southern China, has a tiny world population and no known wintering grounds (though one was found in Cambodia in winter 2012/13). These findings raised the possibility that small numbers of these species could winter further south than previously believed.

Despite its brilliant plumage, Blue-winged Pitta (*Pitta moluccensis*) is unobtrusive as it feeds on the forest floor. This pitta is a passage migrant at Fraser's Hill in autumn and spring.

Swifts, pittas, bee-eaters and thrushes

The influx from the north takes place largely in November. Slaty-legged Crake is never an easy bird to find but a determined – and lucky! – birder prepared to spend time near one of the streams flowing through the forest could be

rewarded with views of this hard-to-see species. It breeds in Burma, northern Thailand, Laos and Vietnam and winters south to Sumatra. Less of a challenge, the incredibly aerial Himalayan Swiftlet is one of a number of fairly obvious passage migrants and winter visitors, arriving in November and leaving for its breeding grounds in the Himalayas and mountains of southern China and northern Thailand in March. Nocturnal migration has been noted in November at Fraser's Hill. Its huge relative, the magnificent Silver-backed Needletail also passes through the area as it migrates between its breeding range in Bangladesh and winter quarters in the south of Malaysia and Sumatra. It is mostly a diurnal migrant between late October and early December in the Cameron Highlands but there have also been records of nocturnal passage. This giant swift moves north in April and May.

While much of the Blue-tailed Bee-eater population is resident in Indochina, birds breeding in southern China and Burma move south in autumn, fanning out through Malaysia and into Sumatra and Java. Some arrive at Fraser's Hill in October, where they spend the winter.

Most pittas are sedentary, but Blue-winged and Hooded Pittas are exceptions, and both species migrate through Fraser's Hill in spring and autumn. Their movements go almost unnoticed, however. Being denizens of the forest floor in low densities, they are easily overlooked. Blue-winged Pittas breed in Indochina and the extreme south of China and move south in September

A few Black-capped Kingfishers (*Halcyon pileata*) appear at Fraser's Hill in autumn as they move south from China and Burma, where they breed, to their winter domiciles in Malaysia and Indonesia.

and October. Their passage through Peninsular Malaysia reaches a peak in the second half of October. Some remain for the winter but most continue south to Sumatra and Borneo; a handful of vagrants have even been seen in north-western Australia. From late March they undertake the return journey north. Hooded Pittas of the subspecies *cucullata* breed in the Himalayan foothills of northern India, parts of southern China and northern Vietnam. They move south to Sumatra (and maybe Borneo) from mid-October to December, flying by night. Return migration is in April and May. Siberian Thrush arrives at around the same time in autumn as Eye-browed Thrush. Indeed, they sometimes migrate together in mixed flocks of up to 50 or 60, and mostly at night. In late March or April they depart for the land that gave them their name.

One of the commonest migrants at Fraser's Hill is Eye-browed Thrush, a species that deserts its Siberian breeding grounds entirely in the autumn to winter in Indochina, Malaysia and Sumatra. Although most have left their summer domicile by early September they do not arrive at the hill station until late November or December, leaving again in March and April.

An adult male Siberian Blue Robin (*Luscinia cyane*), a passage migrant and winter visitor. This magical chat breeds in south-east Siberia, northern China, Korea and Japan. Birds often visit forest pools to bathe in late afternoon.

Grey Nightjar

Grey Nightjar (*Caprimulgus jotaka*) is a long-distance migrant, but not much is known about the routes it follows. It breeds across a vast area of China, the far east of Siberia, Korea and Japan, and winters in Malaysia, Sumatra, Java, Borneo and the Philippines. Spring migration is between March and May, when it can be seen coming in off the sea near Beidaihe, China, in the morning but most probably migrate overland, by night. It leaves its breeding grounds from September to November, when it is seen in forested areas such as Fraser's Hill. Some ornithologists consider it to be a subspecies of Jungle Nightjar (*C. Indicus*), whose other populations are sedentary.

The movements of Orange-headed Thrush are much more complex. This species breeds no further north than southern China, and some populations are sedentary. It is recorded mostly as a late autumn migrant at Fraser's Hill (mainly in late November and December), with only small numbers passing north in April and May. That said, as a bird of thick forest it is much less noticeable than the previous species.

Other autumn arrivals include Oriental Honey-buzzard, which moves south mostly in October, Red-legged Crake, Black-capped Kingfisher, Richard's Pipit,

Orange-headed Thrush (*Zoothera citrina*) is most likely to be encountered in November and December. It is another unobtrusive forest species.

Lanceolated, Eastern Crowned and Arctic Warblers, Mugimaki, Asian Brown and Dark-sided Flycatchers, Tiger and Brown Shrikes and Forest Wagtail, with its characteristic side-to-side tail movement.

Working the area

Entrance to the area is by road, through The Gap, where there is good forest at 800 metres. Migrants and winter visitors can be found here. For those with the stamina, birding uphill along the winding Old Road towards the resort itself is always productive. Since it is 8 kilometres to the top, at more than 1,200 metres, it may be best to walk part of the way and then return. Subtle changes in the avifauna take place with increasing altitude, though these are more apparent among the resident species. When the sun goes down, the Old and New roads between The Gap and the resort are good for nightjars, including Grey. The golf course is worth checking as are the gardens and lawns around the resort's bungalows and hotels. Just north of the golf course is Hemmant Trail, good for many forest species; to the west, Bishop's Trail is longer and basically an extension of this. And for anyone feeling particularly energetic the 6-kilometre Pine Tree Trail is even longer, steeper and more slippery and it also offers unwelcome opportunities to encounter leeches.

The Titiwangsa Range around Fraser's Hill comprises forested ridges, peaks and valleys with many streams and small forest pools – ideal habitats for a wide range of resident and migrant bird species.

AUSTRALIA

Bribie Island

Coral Sea

MORETON ● BAY

QUEENSLAND

Brisbane ●

POINT ○ LOOKOUT

- **LOCATION**
  South-east Queensland, Australia

- **FLYWAY**
  East Asian–Australasian

- **SPRING**
  Waders, including Far-eastern Curlew and Grey-tailed Tattler; seabirds; Yellow-faced Honeyeater, Channel-billed and Australian Golden-bronze Cuckoos, Sacred Kingfisher and Rainbow Bee-eater pass north.

- **AUTUMN**
  Waders; northern-breeding terns, including White-winged Black, return; seabirds, including thousands of Short-tailed Shearwaters offshore in November.

- **KEY SITES**
  Caloundra, Kakadu Beach, Toorbul, Wynnum–Manly, Point Lookout.

- **THREATS**
  Disturbance of wader and tern roosts; coastal development.

Moreton Bay, Australia

Each year up to two million shorebirds migrate to and from Australia, and a good percentage of them use the Moreton Bay area of Queensland either as a staging post or somewhere to spend the austral summer.

Australia does not have raptor bottlenecks like Veracruz or Eilat. And no Australian site can compete with the exciting passerine migration of Point Pelee or Beidaihe. However, what it does have is a chain of intertidal feeding grounds on the east coast, which form the southern end of the East Asian–Australasian Flyway. One of the most important is Moreton Bay, which can host up to 40,000 waders during spring and autumn passage and in the austral summer. On the ocean side of the bay is a series of islands with ocean-facing headlands, such as Point Lookout. These can lay claim to some of the most exciting seawatching in Australia.

Moreton Bay and Pumicestone Passage, which runs into it from the north, are close to Brisbane in south-east Queensland. The bay itself is up to 80 kilometres from north to south and almost completely enclosed by a series of islands, four of the largest sand islands in the world. Long sections of the bay's margins are suitable habitat for waders and other waterbirds. There are saltmarshes and intertidal mudflats and sandflats. These habitats provide feeding and roosting sites for thousands of migratory waders, and many other resident ones. For many of the migrants the area is a terminus, the southernmost point on a migration that has brought them from Siberia and Alaska. For others (Bar-tailed Godwits, for example), from late August through to

Black Noddy (*Anous minutus*) is a small tern species that breeds in offshore colonies and disperses south to waters east of Moreton Bay.

A tree roost of Terek Sandpipers (*Xenus cinereus*) at Glass House Mountain Creek, a site just north of Moreton Bay.

October, it is a staging area, their first Australian port of call (or at least, their second if they have also broken their journey in the Gulf of Carpentaria) en route to south-east Australia or New Zealand. The area is also important on the northward migration of waders between the end of January (Far-eastern Curlew) and April or early May (Grey-tailed Tattler). In fact, it may be their last stop-over before having to cross wide stretches of ocean. Moreton Bay is especially important for Far-eastern Curlew and Grey-tailed Tattler, supporting up to 20 per cent and 50 per cent of the world populations, respectively.

Seabirds

The area's other claim to fame is its seabirds. Although these can sometimes be seen within the bay, the best seawatching is from one of the headlands looking out into the Coral Sea. Cold water from the Southern Ocean meets the warmer waters of the southward-moving East Australia Current off the coast of southern Queensland, creating zones rich in invertebrate prey for hungry shearwaters, petrels, storm-petrels and albatrosses. Several factors combine to produce excellent seawatching, including the exact location of the most food-rich zone, the prevalence of onshore winds pushing them closer to the coast and the time of year.

There is a broad distinction between those seabirds that spend most of the year in the area, those that disperse through the southern South Pacific after breeding further south (and so in a sense are in a state of more-or-less constant migration) and those that move north through the area into tropical waters or the North Pacific when they have finished breeding. One of the best examples of the last is Short-tailed Shearwater, which moves north from its breeding grounds well out in the Pacific but returns much closer to the Queensland coast in the austral spring.

Several high-tide wader roosts around Moreton Bay hold Great Knot (*Calidris tenuirostris*) during the austral summer.

Terns and waders

Many terns from the northern hemisphere overwinter in Australia and one of the highlights of the birding year is the huge mixed flocks of Common, Little, and White-winged Black Terns that build up at Golden Beach, Caloundra, at the northern end of Pumicestone Passage. Numbers build dramatically between late January and mid-February and there may be up to 40,000. Common Tern is the most numerous. Most have left by the end of April.

Also early in the year the bay's migrant shorebirds reach their peak as more southerly birds supplement those already there. Thereafter, numbers fall as they begin their long northbound passage towards China, Korea, Siberia and

Alaska. Far-eastern Curlews are among the first to leave. They will stop off at sites around the Yellow Sea before continuing. Grey-tailed Tattlers are one of the last species of wader to leave; the next stop for many will be Japan.

Between March and May small numbers of migrant passerines and near-passerines may be seen on their way north. Passerine migration is not impressive but small flocks of Yellow-faced Honeyeaters may be on the move north, having bred in south-east Australia. Channel-billed Cuckoos, Sacred Kingfishers and Rainbow Bee-eaters also pass through the area at this time to avoid the less favourable conditions of the southern Australian winter. While northern populations of Sacred Kingfisher are resident, southern birds join them or fly on to New Guinea or Borneo for the austral winter.

Meanwhile, the northern hemisphere waders are replaced by Double-banded Plovers, which breed on pebble beds in rivers on New Zealand's South Island. Some spend the austral winter (March to September) in Moreton Bay, arriving in late March. A small number of White-fronted Terns appear, having bred in New Zealand early in the year.

Seawatching can be fascinating at any time but there are periods when it can be breathtaking: March and April, June and July (best for albatrosses), and October and November (big numbers of shearwaters). Then, in the right conditions, with a deep low pressure system to the north-east producing persistent south-easterlies, coastal headlands can be excellent. Possibilities include Wandering, Black-browed and Yellow-nosed Albatrosses, Providence, Cape and Great-winged Petrels, Fairy Prion and White-fronted Tern. Rarities have included Tahiti, White-necked and Gould's Petrels, Flesh-footed Shearwater and Masked and Red-footed Boobies. The first waders begin to reappear again in mid- to late July, with numbers building to a peak in October. Of these, Bar-tailed Godwits are the most numerous.

Bridled Tern (*Onychoprion anaethetus*) breeds further to the north but is an occasional visitor to Moreton Bay.

Beginning in October, migrant Common, Little and White-winged Black Terns arrive from northern hemisphere breeding grounds. Numbers are generally highest between late December and mid-April. October is the best month for some species of seabirds, notably Kermadec Petrel. Then in November thousands of Short-tailed Shearwaters stream south along the coast towards their Tasmanian breeding grounds.

Toorbul wader roost

The wader afficionado is spoilt for choice: more than 100 wader roost sites have been identified around Moreton Bay. Some are small, others are hard to access and some of those that are easy to reach are subject to human disturbance. However, several stand out and are worth a visit. Check tide times before you visit because – like anywhere – waders disperse across the intertidal zone to feed when the tide is not high.

Starting to the north of the bay, on the west side of Pumicestone Passage the small town of Toorbul has an artificially enhanced roost site along the Esplanade. A little further north is the smaller Toorbul Sandfly Bay roost, which may be used if the main roost is disturbed, and Toorbul Sandspit roost, at the northern end of the Esplanade. Good views of the Toorbul roost are possible when the tide is up and on incoming and outgoing tides. However, if the tide is above 2 metres the birds may be forced to move elsewhere. The roost is less disturbed on weekdays than on weekends. It may be possible to see 50 or so Red Knot among larger numbers of Great Knot here during September and October before they move on to sites further south in Australia or in New Zealand. Bar-tailed Godwit is a common summering species, there is a good chance of seeing some Sharp-tailed Sandpipers and Red-necked Stints are possible on their way north. On very high tides small numbers of Pacific Golden Plover can be found at one of the smaller Toorbul roosts. Scarce or rare waders such as Broad-billed Sandpiper, Grey Plover and Asian Dowitcher have all been seen here.

Caloundra sandflats

A little further north, between the town of Caloundra and the northern end of Bribie Island, are some large intertidal sandflats. Spectacular numbers of terns, in particular migrant species, use these as roost sites. Common, White-winged Black and Little Terns join resident species from late October. Those that continue on their way south are replaced by others, and large numbers are ever-present until April and May when the pattern is repeated in reverse. Detailed studies have produced peak counts of 38,000 Common, 17,000 White-winged Black and 11,000 Little Terns. These are among the

biggest concentrations in Australia; indeed, the Common Tern figure is one of the biggest on the planet. Resident or nomadic Greater Crested, Caspian and Gull-billed Terns are also present. Caloundra can also be a reasonable seawatching location.

Kakadu Beach

On the other side of Pumicestone Passage, Kakadu Beach on Bribie Island is an artificial roost site built to replace one destroyed by development. Although it does not attract the three-figure numbers of Far-eastern Curlew that frequented the old roost (Dux Creek), the species is still hanging on here. Especially on the highest tides the 200-metre-long roost can hold Bar-tailed Godwit, Whimbrel, Great Knot, Red-necked Stint, Curlew Sandpiper, Lesser Sand Plover and Greater Sand Plover. However, it is a bit hit and miss. Kakadu is also a regular site for Double-banded Plover, visiting from New Zealand between March and August.

On the west shore of Moreton Bay, Deception Bay is best reached from the main Bruce highway north of Brisbane. It can be a good site for Red-necked Avocet, a species of ephemeral inland lakes that moves to the coast when

Thousands of Common Terns (*Sterna hirundo*), along with smaller numbers of White-winged Black (*Chlidonias leucopterus*), Little (*Sternula albifrons*) and other species use the Caloundra sandflats in the austral summer. Human disturbance can be a problem for them.

drought conditions prevail inland. In May 2005 there was a record total of 646 there. South again, the foreshore at Wynnum–Manly is a good site for Terek Sandpiper, Great Knot, Red Knot, Ruddy Turnstone, Pacific Golden Plover, Lesser Sand Plover and a few Greater Sand Plover. These birds, and others, can be viewed from the esplanade two or three hours either side of high tide.

Point Lookout

Point Lookout is a headland at the north-east tip of North Stradbroke Island ('Straddy' to the locals). The island is mostly sandy but the headland is a rocky vantage point, which offers some of the best land-based seawatching in Australia. It is pretty much as far east as you can get on the mainland. Looking east, there is nothing between it and the coast of South America apart from a few atolls and thousands of kilometres of South Pacific Ocean. Also good news is the fact that Point Lookout is relatively straightforward to reach. Take

From January to March (depending on where they have bred) Channel-billed Cuckoos (*Scythrops novaehollandiae*) migrate north through eastern Australia to wintering grounds in Papua New Guinea and Indonesia.

the car ferry from Cleveland to Dunwich and drive to the headland; or use the pedestrian ferry and the bus service to the point.

Prime conditions are after protracted south-easterly blows in the austral winter (April to September) and summer cyclones. Seabirds can pass in spectacular numbers. For example, more than 5,000 of both Wedge-tailed and Short-tailed Shearwaters were logged on a single day in October 2010.

Scarcer migrants known to pass this brilliant seawatching location include Providence Petrels and Wilson's Storm-petrels, which move north in May and June, and Buller's Shearwaters, which migrate north between November and January (before returning to their New Zealand breeding grounds via the west coast of South America). Add to the mix Wandering, Black-browed, Yellow-nosed and Buller's Albatrosses; Hutton's, Little, Wedge-tailed, Flesh-footed and Buller's Shearwaters; Cape, Great-winged, Providence, White-necked and Northern Giant Petrels; and Fairy Prion and you have the recipe for classic seawatching without spending hours at sea. There is usually something of interest, but the best months are February and March, June and July, and October and November. For the maximum variety, June is probably best. One weekend in mid-June 2012 the following passing seabirds were noted: one Black-browed Albatross, seven Northern Giant Petrels, six Providence Petrels and 20 Great-winged Petrels; one Fairy Prion; two Sooty and hundreds of Fluttering or Hutton's Shearwaters; one Wilson's Storm-petrel; and two each of Pomarine and Arctic Skuas.

Pelagic alternatives

Seawatching alternatives include Cape Moreton, on the northern tip of Moreton Island (although this is a trickier and more expensive trip), and two regular pelagic trips: the Southport pelagic from the Gold Coast, south of Brisbane, and the Mooloolaba pelagic from the Sunshine Coast north of the city. These have the advantage of getting further out into the ocean and often produce some astounding records. The Southport pelagics run out to Jim's Mountain, a seamount about 50 kilometres north-north-east of the port of embarkation. As of July 2012, 81 species of seabirds had been recorded on the Southport trip, the latest being Light-mantled Sooty Albatross (three on 16 June 2012). Depending on the season, a good variety of albatrosses, petrels and storm-petrels is likely.

Seabirds also appear in Moreton Bay itself. Up to 2,000 Fluttering Shearwaters may be present in the northern section, off Bribie Island, in July and early August, sometimes accompanied by rarer tubenoses such as Hutton's Shearwater. Australasian Gannets are also regular, sometimes numerous, in the austral winter.

References

Alerstam, T. (trans. David Christie). 1993. *Bird Migration*. Cambridge University Press, Cambridge.

Bamford, M. D., Watkins, D., Bancroft, W., Tischler, G. and Wahl, J. 2008. *Migratory Shorebirds of the East Asian–Australasian Flyway; population estimates and internationally important sites.* Canberra, Wetlands International Oceania.

Batumi Raptor Count, www.batumiraptorcount.org

Berthold, P. 2001. *Bird Migration: A General Survey.* Oxford University Press, Oxford.

Birding Cadiz Province, www.birdingcadizprovince.weebly.com

Birding Israel, www.birdingisrael.com

BirdLife International Data Zone, www.birdlife.org/datazone/home

Birdwatch Cork, www.birdwatchcork.com

BirdWatch Ireland, www.birdwatchireland.ie

Birdwatching Bulgaria, www.neophron.com

Brazil, M. 2010. *Birds of East Asia*. Christopher Helm, London.

Chan, K. and Dening, J. 2007. Use of sandbanks by terns in Queensland, Australia; a priority for conservation in a popular recreational waterway. *Biodiversity and Conservation* 16, 447–464.

Cornwall Birding, www.cornwall-birding.co.uk

Corso, A. 2001. Raptor migration across the Strait of Messina, *Brit. Birds* 94: 196–202.

Couzens, D. 2005. *Bird Migration*. New Holland, London.

Dudley, S. 2010. *A Birdwatcher's Guide to Lesvos* (privately published).

Dudley, S. 2010, 2011, 2012. *Lesvos Birds 2009, Lesvos Birds 2010, Lesvos Birds 2011*. Lesvos Birding.

Elkins, N. 2004. *Weather and Bird Behaviour*. T. & A.D. Poyser, London.

Elphick, J. 2007. *The Natural History Museum Atlas of Bird Migration.* Natural History Museum, London.

Estonian Nature Tours, www.naturetours.ee

Falsterbo Bird Observatory, www.falsterbofagelstation.se

Fraser's Hill Birds, www.fraserhill.info/bird-watching.htm

Gätke, H. 1895. *Heligoland as an Ornithological Observatory: the result of fifty years' experience.* Douglas, Edinburgh.

Grimmett, R., Inskipp, C. and Inskipp, T. 2009. *Birds of the Indian Subcontinent.* Christopher Helm, London.

Hawk Mountain Sanctuary, www.hawkmountain.org

HawkWatch International, www.hawkwatch.org
Heligoland Bird Observatory, www.oag-helgoland.de
Hong Kong Birdwatching Society, www.hkbws.org.uk
Houston Audubon Society, www.houstonaudubon.org
Howell, S. 1999. *Where to Watch Birds in Mexico*. Christopher Helm, London.

Isles of Scilly Bird Group, www.scilly-birding.co.uk
Israel Ornithological Center, www.israbirdcenter.org

Leshem, Y. and Yom-Tov, Y. 1997. Routes of migrating soaring birds. *Ibis* 140: 41–52.
Lesvos Birding, http://athene-birdinglesvos.blogspot.co.uk
Long Point Bird Observatory, www.bsc-eoc.org/longpoint/index.jsp?targetpg=index&lang=EN

Malaysia Birding, www.malaysiabirding.org

Ogilvie, M. and Winter, S. (eds.). 1989. *Best Days with British Birds*. British Birds, London.
Ontario Bird Watching, www.birding.com/wheretobird/ontario.asp

Point Pelee National Parks of Canada (Natural Wonders and Cultural Treasures),
    www.pc.gc.ca/eng/pn-np/on/pelee/natcul.aspx

Queensland Wader Study Group, www.waders.org.au

Ritsema, A. 2007. *Heligoland Past and Present*. Lulu Publishing, Raleigh.
Robson, C. 2009. *Birds of South-East Asia*. New Holland, London.

Sibley, D. 2000. *The North American Bird Guide*. Pica Press, Robertsbridge.
Slack, R. 2009. *Rare Birds Where and When (Sandgrouse to New World Orioles)*. Rare Bird Books, York.
Snow, D. W. and Perrins, C. M. 1998. *The Birds of the Western Palearctic,*
    concise edition, volumes 1 and 2. Oxford University Press, Oxford.
Spurn Bird Observatory, www.spurnobservatory.co.uk
Svensson, L., Grant, P. J., Mullarney, K., Zetterstom, D. 2010. *Collins Bird Guide*.
    Harper Collins, London.

Thai Raptor Group, www.thairaptorgroup.com
The Birds of Kazakhstan, http://birdsofkazakhstan.com

Wheatley, N. 1996. *Where to Watch Birds in Asia*. Christopher Helm, London.
World Birding Center, Texas, www.theworldbirdingcenter.com

Zimmerman, D., Turner, D., Pearson, D., Willis, I., Pratt, H. D. 2005. *Birds of Kenya and Northern
    Tanzania*. Christopher Helm, London.

Acknowledgements

I would like to thank Bloomsbury's Jim Martin for arguing the case for what was once just an idea, and Julie Bailey for guiding the book's progress. Thanks are also due to Barrie Coo at the RSPB for looking over the manuscript and to Marianne Taylor for her proofreading. *Migration Hotspots* would not work without its superb photographs, so I am very grateful to those who offered their images. Special praise should go to Neil Bowman and Stefan Oscarsson, who between them have contributed about half the book's pictures – and who have also been excellent travelling companions over the years. I would also like to take this opportunity to thank the ornithological legend Chris Perrins for agreeing to write the foreword and the following for checking texts on specific locations and offering other assistance: Mark Andrews; Susan Billetdeaux and Mary Anne Weber (Houston Audubon Society); Neil Bowman; John Cantelo; Robert DeCandido; Linda Cross; Jill Dening; Jochen Dierschke (Heligoland Bird Observatory); Steve Dudley (Lesvos Birding, http://athene-birdinglesvos.blogspot.co.uk); Trevor Ford; Dimiter Georgiev (Neophron, www.neophron.com); Anna Giordano and Simonetta Cutini; Gerard Gorman; Jerry and Linda Harrison (Canopy Tower Family, www.canopytower.com); Phil Heath; Johannes Jansen, Brecht Verhelst and Wouter Vansteeland (Batumi Raptor Count, www.batumiraptorcount.org); Marika Mann and Tarvo Valker (Estonian Nature Tours, www.naturetours.ee); Elisa Peresbarbosa Rojas and Eduardo Martinez (Pronatura Veracruz, pronaturaveracruz.org); Stefan Oscarsson; Colin Reid; Itai Shanni (Israeli Ornithological Centre, www.israbirdcenter.org); Bena Smith; Rich Stallcup and Lishka Arata (Point Reyes Bird Observatory); Jugal Tiwari; Martin Williams; Malcolm Wilson; Stuart Winter; Lee Zieger and Roberta Jackson (South Padre Island Birding and Nature Center, spibirding. com). Finally, a big thank you to my wife Sharon, who had to put up with my nightly disappearing act as I secreted myself away with the laptop to work on the text.

Photograph credits

Bloomsbury would like to thank the following for providing photographs and for permission to reproduce copyright material. While every effort has been made to trace and acknowledge all copyright holders, we would like to apologise for any errors or omissions and invite readers to inform us so that corrections can be made in any future editions of the book.

Deborah Allen 198; Mark Andrews 12, 191; Robert DeCandido 196, 197; Neil Bowman title, 4, 10, 14, 16, 77, 79, 81, 82, 84, 88, 89, 95, 98, 102, 106, 119, 120, 123, 125, 126, 127, 128, 129, 132, 133, 137, 144, 148, 151, 159, 160, 161, 163, 164, 165, 166, 167, 168, 169, 171, 172, 173, 174, 175, 176, 178, 180, 181, 183, 184, 187, 188, 189, 195, 200, 202, 203, 204, 206, 208, 209, 212; John Cantelo 107; Jill Dening 207, 211; Steve Dudley 131, 135; Neil Fifer 186, 191, 194; Christian Gelpke/Batumi Raptor Group 138, 139; Anna Giordano 113; Tim Harris 117; Houston Audubon Society 43; John Jackson 96; Kaarel Kaisel/Estonian Nature Tours 8; Jonathan Lethbridge (JustBirdPhotos.com) 91, 92, 149, 150, 152; Mariano Martinez Lopez 112; naturespicsonline.com 21, 22, 49, 50; Christian Nielsen 136; Stefan Oscarsson 63, 65, 66, 68, 69, 71, 74, 76, 85, 87, 99, 105, 108, 114, 143, 147, 155; James Packer 134; Jari Peltomaki/Estonian Nature Tours 74; Public Domain: 20, 29, 32, 46, 59, 78, Chris Beckwith 48, Cephas 41, JJ Harrison 100, Minette Layne 23, Tim Lenz 44, Vinod Paneker 179, RSPB 101; Remo Savisaar/Estonian Nature Tours 72; Roberto Santoro 115; Shutterstock: 11, 56, 146, Florian Andronache 15, Steve Byland 28, Gerald A. DeBoer 34, 47, Dreamframer 60, Niall Dunne 157, Melinda Fawver 55, Florida Stock 31, 57, Aki Jinn 205, Aleksander Kaasik 70, M Lorenz 39, Anatolly Lukich 33, S. Mishonja 121, Alexei Novikov 64, Tom Reichner 52, Stubblefield Photography 35, Ira Torlin 192; Thinkstock: istockphoto 19, 38, 53, 116, 185, Top Photo Group 146, 182; Freek Verdonckt/Batumi Raptor Group 140.

Index